Universitext

Editors

F.W. Gehring
P.R. Halmos
C.C. Moore

Universitext

Editors: F.W. Gehring, P.R. Halmos, C.C. Moore

Booss/Bleecker: Topology and Analysis
Chern: Complex Manifolds Without Potential Theory
Chorin/Marsden: A Mathematical Introduction to Fluid Mechanics
Cohn: A Classical Invitation to Algebraic Numbers and Class Fields
Curtis: Matrix Groups, 2nd ed.
van Dalen: Logic and Structure
Devlin: Fundamentals of Contemporary Set Theory
Edwards: A Formal Background to Mathematics I a/b
Edwards: A Formal Background to Higher Mathematics II a/b
Endler: Valuation Theory
Frauenthal: Mathematical Modeling in Epidemiology
Gardiner: A First Course in Group Theory
Godbillon: Dynamical Systems on Surfaces
Greub: Multilinear Algebra
Hermes: Introduction to Mathematical Logic
Hurwitz/Kritikos: Lectures on Number Theory
Kelly/Matthews: The Non-Euclidean, The Hyperbolic Plane
Kostrikin: Introduction to Algebra
Luecking/Rubel: Complex Analysis: A Functional Analysis Approach
Lu: Singularity Theory and an Introduction to Catastrophe Theory
Marcus: Number Fields
McCarthy: Introduction to Arithmetical Functions
Meyer: Essential Mathematics for Applied Fields
Moise: Introductory Problem Course in Analysis and Topology
Øksendal: Stochastic Differential Equations
Porter/Woods: Extensions of Hausdorff Spaces
Rees: Notes on Geometry
Reisel: Elementary Theory of Metric Spaces
Rey: Introduction to Robust and Quasi-Robust Statistical Methods
Rickart: Natural Function Algebras
Schreiber: Differential Forms
Smoryński: Self-Reference and Modal Logic
Stanisić: The Mathematical Theory of Turbulence
Stroock: An Introduction to the Theory of Large Deviations
Tolle: Optimization Methods

Paul J. McCarthy

Introduction to Arithmetical Functions

Springer-Verlag
New York Berlin Heidelberg Tokyo

Paul J. McCarthy
Department of Mathematics
University of Kansas
Lawrence, KS 66045
U.S.A.

AMS Classifications: 10-01, 10A20, 10A21, 10H25

Library of Congress Cataloging-in-Publication Data
McCarthy, Paul J. (Paul Joseph)
Introduction to arithmetical functions.
(Universitext)
Bibliography: p.
Includes index.
1. Arithmetic functions. I. Title.
QA245.M36 1985 512′.7 85-26068

With 6 illustrations.

Printed and bound by R.R. Donnelley & Sons, Harrisonburg, Virginia.
Printed in the United States of America.

9 8 7 6 5 4 3 2 1

ISBN 0-387-96262-X Springer-Verlag New York Berlin Heidelberg Tokyo
ISBN 3-540-96262-X Springer-Verlag Berlin Heidelberg New York Tokyo

Preface

The theory of arithmetical functions has always been one of the more active parts of the theory of numbers. The large number of papers in the bibliography, most of which were written in the last forty years, attests to its popularity. Most textbooks on the theory of numbers contain some information on arithmetical functions, usually results which are classical. My purpose is to carry the reader beyond the point at which the textbooks abandon the subject. In each chapter there are some results which can be described as contemporary, and in some chapters this is true of almost all the material.

This is an introduction to the subject, not a treatise. It should not be expected that it covers every topic in the theory of arithmetical functions. The bibliography is a list of papers related to the topics that are covered, and it is at least a good approximation to a complete list within the limits I have set for myself. In the case of some of the topics omitted from or slighted in the book, I cite expository papers on those topics.

Each chapter is followed by notes which are bibliographical in nature, and only incidentally historical. My purpose in the notes is to point out sources of results. Number theory, and the theory of arithmetical functions in particular, is rife with rediscovery, so I hope the reader will not be too harsh with me if I fail to pin down the truly first source of some result. Perhaps this book will help reduce the rate of rediscovery.

There are more than four hundred exercises. They form an essential part of my development of the subject, and during any serious reading of the book some time must be spent thinking about the exercises.

I assume that the reader is familiar with calculus, including infinite series, and has the maturity gained from completing several mathematics courses at the college level. A first course in the theory of numbers provides more than enough background in number theory. In fact, only a few things from such a course are used, such as elementary properties of congruences and the unique factorization theorem.

Table of Contents

Chapter 1. Multiplicative Functions 1

Chapter 2. Ramanujan Sums 70

Chapter 3. Counting Solutions of Congruences 114

Chapter 4. Generalizations of Dirichlet Convolution 149

Chapter 5. Dirichlet Series and Generating Functions 184

Chapter 6. Asymptotic Properties of Arithmetical Functions 255

Chapter 7. Generalized Arithmetical Functions 293

References 333

Bibliography 334

Index 361

Chapter 1
Multiplicative Functions

Throughout this book r and n, and certain other letters, are integer variables. Without exception r is restricted to the positive integers. Unless it is stated to the contrary, the same is true of n and other integer variables. However, on some occasions n and other integer variables will be allowed to have negative and zero values. On such occasions it will be stated explicitly that this is the case.

An <u>arithmetical function</u> is a complex-valued function defined on the set of positive integers.

Although examples of such functions can be defined in a completely arbitrary manner, the most interesting ones are those that arise from some arithmetical consideration. Our first examples certainly do arise in this way. The <u>Euler function</u> ϕ is defined by

$$\phi(n) = \text{the number of integers } x \text{ such that } 1 \le x \le n \text{ and } (x, n) = 1.$$

If k is a nonnegative integer, the function σ_k is defined by

$$\sigma_k(n) = \text{the sum of the } k\text{th powers of the divisors of } n .$$

The term "divisor" always means "positive divisor." We can express $\sigma_k(n)$ by using the sigma-notation:

$$\sigma_k(n) = \sum_{d|n} d^k \quad \text{for all } n .$$

In particular,

$$\sigma(n) = \sigma_1(n) = \text{the sum of the divisors of } n ,$$

and

$$\tau(n) = \sigma_0(n) = \text{the number of divisors of } n .$$

If k is a nonnegative integer, the function ζ_k is defined by $\zeta_k(n) = n^k$. The function $\zeta = \zeta_0$ is called the zeta function:

$$\zeta(n) = 1 \quad \text{for all } n .$$

There are several useful binary operations on the set of arithmetical functions. If f and g are arithmetical functions their sum $f + g$ and product fg are defined in the usual way:

$$(f + g)(n) = f(n)+g(n) \qquad \text{for all } n$$

and

$$(fg)(n) = f(n)g(n) \qquad \text{for all } n .$$

The (Dirichlet) convolution $f * g$ of f and g is defined by

$$(f * g)(n) = \sum_{d|n} f(d)g(n/d) \qquad \text{for all } n .$$

For example, $\sigma_k = \zeta_k * \zeta$.

Addition and multiplication of arithmetical functions have all the usual properties of commutativity, associativity, distributivity, etc.

Proposition 1.1. If f, g and h are arithmetical functions then

(i) $f * g = g * f$,

(ii) $(f * g) * h = f * (g * h)$,

(iii) $f * (g + h) = f * g + f * h$.

Proof. (i) follows from the fact that as d runs over all the divisors of n, so does n/d. Now, for all n,

$$(f * (g * h))(n) = \sum_{d|n} f(d) \sum_{e|\frac{n}{d}} g(e)h(n/de) ,$$

and if $D = de$ this is equal to

$$\sum_{D|n} \Big(\sum_{e|D} f(D/e)g(e)\Big)h(n/D)$$

$$= ((f * g) * h)(n).$$

This proves (ii). As for (iii), for all n,

$$(f * (g + h))(n) = \sum_{d|n} f(d)(g(n/d) + h(n/d))$$

$$= \sum_{d|n} f(d)g(n/d) + \sum_{d|n} f(d)h(n/d)$$

$$= (f * g)(n) + (f * h)(n)$$

$$= (f * g + f * h)(n). \quad \square$$

Thus, in the language of abstract algebra, the set of arithmetical functions, together with the binary operations of addition and convolution, is a commutative ring $\mathfrak{A}$. The ring $\mathfrak{A}$ has a unity. If the arithmetical function δ is defined by

$$\delta(n) = \begin{cases} 1 & \text{if} \quad n = 1 \\ 0 & \text{if} \quad n \neq 1 , \end{cases}$$

then $f * \delta = \delta * f = f$ for all arithmetical functions f.

Let f be an arithmetical function. An arithmetical function g is called an <u>inverse</u> of f if $f * g = g * f = \delta$. If g and g' both have this property then

$$\begin{aligned} g = \delta * g &= (g' * f) * g \\ &= g' * (f * g) = g' * \delta = g' . \end{aligned}$$

Thus, if f has an inverse, it has a unique inverse, which will be denoted by f^{-1}. The elements of the ring $\mathfrak{A}$ that have inverses are its units.

<u>Proposition 1.2</u>. An arithmetical function f has an inverse if and only if $f(1) \neq 0$.

Proof. Suppose that f has an inverse. Then

$$1 = \delta(1) = (f * f^{-1})(1) = f(1)f^{-1}(1) :$$

hence $f(1) \neq 0$. On the other hand, if $f(1) \neq 0$, and if we define the function g inductively by

$$g(1) = \frac{1}{f(1)}$$

and

$$g(n) = -\frac{1}{f(1)} \sum_{\substack{d|n \\ d>1}} f(d)g(n/d) \quad \text{for all } n > 1 \ ,$$

then $f * g = \delta$, and $g * f = \delta$ by Proposition 1.1: g is the inverse of f . □

By this proposition the zeta function ζ has an inverse. It is denoted by μ and called the <u>Möbius function</u>. Since $\mu * \zeta = \delta$, the defining property of μ is

$$\sum_{d|n} \mu(d) = \begin{cases} 1 & \text{if } n = 1 \\ 0 & \text{if } n \neq 1 \end{cases}$$

In particular, if p is a prime and $\alpha \geq 1$,

$$\sum_{j=0}^{\alpha} \mu(p^{\alpha}) = 0 \ .$$

Thus, $\mu(1) = 1$, $\mu(p) = -1$ and $\mu(p^{\alpha}) = 0$ for all $\alpha \geq 2$.

If f and g are arithmetical functions and if $f = g * \zeta$, then $g = f * \mu$. This is the classical <u>Möbius inversion formula</u>. The converse is also true, of course, since ζ and μ are inverses of one another, i.e., if $g = f * \mu$ then $f = g * \zeta$. These statements have a more dramatic visual impact if they are written out using the sigma-notation:

<u>Theorem 1.3</u>. If f and g are arithmetical functions then

$$f(n) = \sum_{d|n} g(d) \quad \text{for all } n$$

if and only if

$$g(n) = \sum_{d|n} f(d)\,\mu(n/d) \quad \text{for all } n\,.$$

As an example, since

$$\sigma_k(n) = \sum_{d|n} d^k \quad \text{for all } n\,,$$

we have

$$n^k = \sum_{d|n} \sigma_k(d)\,\mu(n/d) \quad \text{for all } n\,.$$

We can use Theorem 1.3 to find a formula for $\phi(n)$.

<u>Lemma 1.4</u>. If $d|n$ let

$$S_d = \{xn/d : 1 \leq x \leq d \text{ and } (x, d) = 1\}.$$

If $d, e|n$ and $d \neq e$ then $S_d \cap S_e$ is empty, and

$$\bigcup_{d|n} S_d = \{1, \ldots, n\}\,.$$

Proof. Suppose $S_d \cap S_e$ is not empty. Then there exist x and y such that $1 \leq x \leq d$, $1 \leq y \leq e$, $(x, d) = 1 = (y, e)$ and $xn/d = yn/e$, i.e., $xe = yd$. Since $(x, d) = 1$, $x|y$, and likewise $y|x$. Thus $x = y$ and $d = e$.

If $1 \leq m \leq n$ and $(m, n) = n/d$, and if $m = xn/d$, then $(x, d) = 1$ and $1 \leq x \leq dm/n \leq d$. Thus $m \in S_d$. □

Since S_d has $\phi(d)$ elements,

$$n = \sum_{d|n} \phi(d) \quad \text{for all } n .$$

Therefore, by Theorem 1.3,

$$\phi(n) = \sum_{d|n} d\mu(n/d) \quad \text{for all } n .$$

In particular, if p is a prime and $\alpha \geq 1$,

$$\phi(p^\alpha) = \sum_{j=0}^{\alpha} p^j \mu(p^{\alpha-j})$$

$$= p^\alpha - p^{\alpha-1} = p^\alpha(1 - \frac{1}{p}) .$$

An arithmetical function f is called a <u>multiplicative function</u> if $f(n) \neq 0$ for at least one integer n , and if

$$f(mn) = f(m)f(n) \quad \text{for all } m \text{ and } n \text{ with } (m, n) = 1 .$$

If f is a multiplicative function and if $f(n) \neq 0$, then $f(n) = f(1n) = f(1)f(n)$. Thus, $f(1) \neq 0$ and consequently f has an inverse. Note that $f(1) = 1$. If n is an arbitrary positive integer and if $n = p_1^{\alpha_1} \ldots p_t^{\alpha_t}$, then

$$f(n) = \prod_{i=1}^{t} f(p_i^{\alpha_i}) .$$

Thus, a multiplicative function is completely determined by its values at the prime powers.

Proposition 1.5. If f is a multiplicative function then f^{-1} is a multiplicative function.

Proof. Suppose $(m, n) = 1$. If $m = n = 1$ then $f^{-1}(mn) = 1 = f^{-1}(m)f^{-1}(n)$. Suppose that $mn \neq 1$ and that $f^{-1}(m_1n_1) = f^{-1}(m_1)f^{-1}(n_1)$ whenever $(m_1, n_1) = 1$ and $m_1n_1 < mn$. If either $m = 1$ or $n = 1$ then certainly $f^{-1}(mn) = f^{-1}(m)f^{-1}(n)$. Suppose $m \neq 1 \neq n$.

From the proof of Proposition 1.2,

$$f^{-1}(mn) = - \sum_{\substack{d|mn \\ d>1}} f(d)f^{-1}(mn/d) .$$

Since $(m, n) = 1$, every divisor d of mn can be written uniquely as $d = d_1d_2$, where $d_1|m$ and $d_2|n$: of course, $(d_1, d_2) = 1$ and

$(m/d_1, n/d_2) = 1$. Then

$$f^{-1}(mn) = - \sum_{\substack{d_1|m \\ d_2|n \\ d_1d_2>1}} f(d_1d_2)f^{-1}(mn/d_1d_2) ,$$

and since $(m/d_1)(n/d_2) < mn$ this is equal to

$$- \sum_{\substack{d_1|m \\ d_2|n \\ d_1d_2>1}} f(d_1)f(d_2)f^{-1}(m/d_1)f^{-1}(n/d_2)$$

$$= - f^{-1}(m) \sum_{\substack{d_2|n \\ d_2>1}} f(d_2)f^{-1}(n/d_2) - f^{-1}(n) \sum_{\substack{d_1|m \\ d_1>1}} f(d_1)f^{-1}(m/d_1)$$

$$- (- \sum_{\substack{d_1|m \\ d_1>1}} f(d_1)f^{-1}(m/d_1))(- \sum_{\substack{d_2|n \\ d_2>1}} f(d_2)f^{-1}(n/d_2))$$

$$= f^{-1}(m)f^{-1}(n) + f^{-1}(m)f^{-1}(n) - f^{-1}(m)f^{-1}(n)$$

$$= f^{-1}(m)f^{-1}(n) . \square$$

Since the zeta function ζ is multiplicative, its inverse, the Möbius function μ is multiplicative. Therefore,

$$\mu(n) = \begin{cases} 1 & \text{if } n = 1 \\ (-1)^t & \text{if } n \text{ is the product of } t \text{ distinct primes} \\ 0 & \text{otherwise .} \end{cases}$$

Proposition 1.6. If f and g are multiplicative functions then $f * g$ is multiplicative.

Proof. $(f * g)(1) = f(1)g(1) = 1$, and if $(m, n) = 1$ then

$$(f * g)(mn) = \sum_{d|mn} f(d)g(mn/d)$$

$$= \sum_{\substack{d_1|m \\ d_2|n}} f(d_1)f(d_2)g(m/d_1)g(n/d_2)$$

$$= \Big(\sum_{d_1|m} f(d_1)g(m/d_1)\Big)\Big(\sum_{d_2|n} f(d_2)g(n/d_2)\Big)$$

$$= [(f * g)(m)][(f * g)(n)] \ . \ \Box$$

Since ζ_k is multiplicative and $\sigma_k = \zeta_k * \zeta$, the latter is multiplicative. If $k \geq 1$ and if p is a prime and $\alpha \geq 1$, then

$$\sigma_k(p^\alpha) = \sum_{j=0}^{\alpha} p^{jk} = \frac{p^{(\alpha+1)k} - 1}{p^k - 1}$$

Thus, if $n = p_1^{\alpha_1} \ldots p_t^{\alpha_t}$ then

$$\sigma_k(n) = \prod_{i=1}^{t} \frac{p_i^{(\alpha_i+1)k} - 1}{p_i^k - 1}$$

In particular,

$$\sigma(n) = \prod_{i=0}^{t} \frac{p_i^{\alpha_i+1} - 1}{p_i - 1}$$

Since $\tau(p^{\alpha}) = \alpha + 1$,

$$\tau(n) = \prod_{i=1}^{t} (\alpha_i + 1) .$$

Euler's function ϕ is multiplicative since $\phi = \zeta_1 * \mu$. Thus, for all n ,

$$\phi(n) = n \prod_{p|n} (1 - \frac{1}{p}) ,$$

where the product is over all primes p that divide n .

The function ϕ is a counting function, i.e., for each n , $\phi(n)$ is the number of elements of some set, namely $\{x : 1 \leq x \leq n \text{ and } (x,n) = 1\}$. Counting functions can sometimes be evaluated by using the

Inclusion-Exclusion Principle. If $A_1, \ldots, A_t$ are subsets of a finite set S then

$$\#(S \setminus (A_1 \cup \ldots \cup A_t))$$

$$= \#S + \sum_{j=1}^{t} (-1)^j \sum_{1 \le i_1 < i_2 < \ldots < i_j \le t} \#(A_{i_1} \cap \ldots \cap A_{i_j}).$$

Proof. The proof will be by induction on the number of subsets. The assertion is certainly true when $t = 1$. Suppose that $t > 1$ and that the assertion is true for $t - 1$ subsets. Then

$$\#(S \setminus (A_1 \cup \ldots \cup A_{t-1}))$$

$$= \#S + \sum_{j=1}^{t-1} (-1)^j \sum_{1 \le i_1 < i_2 < \ldots < i_j \le t-1} \#(A_{i_1} \cap \ldots \cap A_{i_j}).$$

Now consider A_t and its $t - 1$ subsets $A_1 \cap A_t, \ldots, A_{t-1} \cap A_t$. Then

$$\#((A_t \setminus (A_1 \cup \ldots \cup A_{t-1})) \cap A_t)$$

$$= \#(A_t \setminus ((A_1 \cap A_t) \cup \ldots \cup (A_{t-1} \cap A_t)))$$

$$= \#A_t + \sum_{j=1}^{t-1} (-1)^j \sum_{1 \le i_1 < i_2 < \ldots < i_j \le t-1} \#(A_{i_1} \cap \ldots \cap A_{i_j} \cap A_t).$$

Thus, the formula for $\#(S \setminus (A_1 \cup \ldots \cup A_t))$ follows from the observation that

$$\#(S \setminus (A_1 \cup \ldots \cup A_t))$$

$$= \#(S \setminus (A_1 \cup \ldots \cup A_{t-1})) - \#((A_t \setminus (A_1 \cup \ldots \cup A_{t-1})) \cap A_t). \quad \square$$

To illustrate the use of this principle we shall evaluate a function that is one of many generalizations of the function ϕ. If k is a positive integer the <u>Jordan function</u> J_k is defined by

$J_k(n)$ = the number of ordered k-tuples of integers $\langle x_1, \ldots, x_k \rangle$ such that $1 \leq x_i \leq n$ for $i = 1, \ldots, k$ and $(x_1, \ldots, x_k, n) = 1$.

Then $J_1 = \phi$. Let $n = p_1^{\alpha_1} \ldots p_t^{\alpha_t}$. If S is the set of all ordered k-tuples of integers $\langle x_1, \ldots, x_k \rangle$ such that $1 \leq x_i \leq n$ for $i = 1, \ldots, k$, and if A_i is the set of all such k-tuples for which $p_i \mid (x_1, \ldots, x_k)$, then

$$J_k(n) = \#(S \setminus (A_1 \cup \ldots \cup A_t)),$$

and

$$\#(A_{i_1} \cap \ldots \cap A_{i_j}) = (n/p_{i_1} \ldots p_{i_j})^k .$$

Thus,

$$J_k(n) = n^k + \sum_{j=1}^{t} (-1)^j \sum_{1 \leq i_1 < i_2 < \ldots < i_j \leq t} \left(\frac{n}{p_{i_1} \cdots p_{i_j}}\right)^k$$

$$= \sum_{d \mid n} (\tfrac{n}{d})^k \mu(d) = \sum_{d \mid n} d^k \mu(n/d) .$$

Therefore, $J_k = \zeta_k * \mu$, and it follows that J_k is a multiplicative function. If p is a prime and $\alpha \geq 1$,

$$J_k(p^\alpha) = p^{\alpha k} - p^{(\alpha-1)k} = p^{\alpha k}\left(1 - \frac{1}{p^k}\right) .$$

Thus,

$$J_k(n) = n^k \prod_{p|n} \left(1 - \frac{1}{p^k}\right) .$$

Another generalization of the Euler function ϕ is defined as follows. If k is a positive integer the <u>von Sterneck function</u> H_k is defined by

$$H_k(n) = \sum_{[e_1,\ldots,e_k]=n} \phi(e_1) \ldots \phi(e_k) ,$$

where the sum is over all ordered k-tuples of integers $\langle e_1, \ldots, e_k\rangle$ such that $1 \le e_i \le n$ for $i = 1, \ldots, k$ and $[e_1, \ldots, e_k] = n$. Note that $H_1 = \phi$ and that $H_k(1) = 1$. Suppose that $(m,n) = 1$ and that $[e_1, \ldots, e_k] = mn$. For $i = 1, \ldots, k$, e_i can be factored uniquely as $e_i = c_i d_i$ where $c_i|m$ and $d_i|n$: then $[c_1, \ldots, c_k] = m$ and $[d_1, \ldots, d_k] = n$. Thus,

$$\begin{aligned} H_k(mn) &= \sum_{[e_1,\ldots,e_k]=mn} \phi(e_1) \ldots \phi(e_k) \\ &= \sum_{\substack{[c_1,\ldots,c_k]=m \\ [d_1,\ldots,d_k]=n}} \phi(c_1) \ldots \phi(c_k)\phi(d_1) \ldots \phi(d_k) \\ &= \sum_{[c_1,\ldots,c_k]=m} \phi(c_1) \ldots \phi(c_k) \sum_{[d_1,\ldots d_k]=n} \phi(d_1) \ldots \phi(d_k) \\ &= H_k(m)H_k(n) . \end{aligned}$$

Therefore, H_k is a multiplicative function. The remarkable thing is that the functions H_k and J_k are the same.

Proposition 1.7. For all k, $H_k = J_k$.

Proof. The proof will be by induction on k. We have noted already that $H_1 = \phi = J_1$. Suppose that $k > 1$ and that $H_{k-1} = J_{k-1}$. Since H_k and J_k are multiplicative functions it suffices to show that if p is a prime and $\alpha \geq 1$ then $H_k(p^\alpha) = J_k(p^\alpha)$. By the induction assumption $H_{k-1}(p^\alpha) = J_{k-1}(p^\alpha)$. Then

$$H_k(p^\alpha) = \sum_{\max(\beta_1,\ldots,\beta_k)=\alpha} \phi(p^{\beta_1}) \ldots \phi(p^{\beta_k})$$

$$= \sum_{\substack{\max(\beta_1,\ldots,\beta_{k-1})=\alpha \\ \beta_k<\alpha}} \phi(p^{\beta_1}) \ldots \phi(p^{\beta_{k-1}})\phi(p^{\beta_k})$$

$$+ \sum_{\max(\beta_1,\ldots,\beta_{k-1})\leq\alpha} \phi(p^{\beta_1}) \ldots \phi(p^{\beta_{k-1}})\phi(p^\alpha)$$

$$= H_{k-1}(p^\alpha) \sum_{d|p^{\alpha-1}} \phi(d) + \left(\sum_{d|p^\alpha} \phi(d)\right)^{k-1} \phi(p^\alpha)$$

$$= p^{\alpha-1} H_{k-1}(p^\alpha) + p^{\alpha(k-1)}\phi(p^\alpha)$$

$$= p^{\alpha-1} J_{k-1}(p^\alpha) + p^{\alpha(k-1)}\phi(p^\alpha)$$

$$= p^{\alpha-1} p^{\alpha(k-1)} (1 - \frac{1}{p^{k-1}}) + p^{\alpha(k-1)} p^\alpha(1 - \frac{1}{p})$$

$$= p^{\alpha k}\left[\frac{1}{p} (1 - \frac{1}{p^{k-1}}) + (1 - \frac{1}{p})\right]$$

$$= p^{\alpha k} (1 - \frac{1}{p^k}) = J_k(p^\alpha) . \square$$

An arithmetical function f is called a <u>completely multiplicative function</u> if $f(n) \neq 0$ for at least one integer n and if

$$f(mn) = f(m)f(n) \quad \text{for all } m \text{ and } n .$$

If f is completely multiplicative then it is, of course, multiplicative. If p is a prime and $\alpha \geq 1$ then $f(p^\alpha) = f(p)^\alpha$. Thus, f is completely determined by its values at the primes.

Among the multiplicative functions the completely multiplicative functions can be distinguished by certain algebraic properties which they possess.

<u>Proposition 1.8</u>. A multiplicative function f is completely multiplicative if and only if $f^{-1} = \mu f$.

Proof. If f is completely multiplicative then for all n ,

$$\begin{aligned}(\mu f * f)(n) &= \sum_{d|n} \mu(d)f(d)f(n/d) \\ &= f(n) \sum_{d|n} \mu(d) \\ &= \begin{cases} f(1) = 1 & \text{if } n = 1 \\ 0 & \text{if } n \neq 1 : \end{cases}\end{aligned}$$

hence $\mu f * f = \delta$.

Conversely, suppose that $f^{-1} = \mu f$. We shall show by induction on α that if p is a prime and $\alpha \geq 1$ then $f(p^\alpha) = f(p)^\alpha$. This is certainly true for $\alpha = 1$. Assume $\alpha \geq 2$ and that $f(p^{\alpha-1}) = f(p)^{\alpha-1}$. For all $\beta \geq 2$,

$$f^{-1}(p^{\beta}) = \mu(p^{\beta})f(p^{\beta}) = 0 ,$$

and from this it follows that

$$0 = (f^{-1} * f)(p^{\alpha}) = f(p^{\alpha}) + f^{-1}(p)f(p^{\alpha-1})$$

$$= f(p^{\alpha}) + f^{-1}(p)f(p)^{\alpha-1}$$

Since $f^{-1}(p) = -f(p)$ (this holds, in fact, for all multiplicative functions),

$$f(p^{\alpha}) = f(p)^{\alpha} . \square$$

From this proof we can extract the following result.

Proposition 1.9. A multiplicative function f is completely multiplicative if and only if $f^{-1}(p^{\alpha}) = 0$ for all primes p and all $a \geq 2$.

Proposition 1.10. A multiplicative function f is completely multiplicative if and only if

$f(g * h) = fg * fh$ for all arithmetical functions g and h .

Proof. If f is completely multiplicative then for all g and h and all n ,

$$(f(g * h))(n) = f(n) \sum_{d|n} g(d)h(n/d)$$

$$= \sum_{d|n} f(d)g(d)f(n/d)h(n/d)$$

$$= (fg * fh)(n) .$$

Conversely, suppose that the equation holds when $g = \zeta$ and $h = \mu$. Then

$$\delta = f\delta = f(\zeta * \mu) = f\zeta * f\mu = f * \mu f ,$$

i.e., $f^{-1} = \mu f$. Therefore, f is completely multiplicative. □

Thus, there are two specific functions g and h, namely $g = \zeta$ and $h = \mu$, such that if f is a multiplicative function and if $f(g * h) = fg * fh$, then f is completely multiplicative. Another pair of functions with this property is given in Exercise 1.44, and two more characterizations of the completely multiplicative functions among the multiplicative functions make up Exercises 1.45 and 1.46.

For each nonnegative integer k the function ζ_k is completely multiplicative. Thus, each of the functions $\sigma_k = \zeta_k * \zeta$ is the convolution of two completely multiplicative functions. Arithmetical functions having this property can be characterized by a condition analogous to the one in Proposition 1.9.

Proposition 1.11. A multiplicative function f is the convolution of two completely multiplicative functions if and only if $f^{-1}(p^\alpha) = 0$ for all primes p and all $\alpha \geq 3$.

Proof. Suppose $f = g * h$ where g and h are completely multiplicative. If p is a prime and $\alpha \geq 3$, then

$$f^{-1}(p^\alpha) = (g^{-1} * h^{-1})(p^\alpha) = \sum_{j=0}^{\alpha} g^{-1}(p^j)h^{-1}(p^{\alpha-j})$$

$= g^{-1}(1)h^{-1}(p^{\alpha}) + g^{-1}(p)h^{-1}(p^{\alpha-1})$ since g is completely multiplicative

$= 0$ since h is completely multiplicative and $\alpha \geq 3$.

Conversely, let f be a multiplicative function for which the condition holds. Let g be a completely multiplicative function such that for each prime p , $g(p)$ is a root of the quadratic equation

$$X^2 + f^{-1}(p)X + f^{-1}(p^2) = 0 .$$

Let $h = g^{-1} * f$. Then h is a multiplicative function and for all primes p and all $\alpha \geq 2$,

$$h^{-1}(p^{\alpha}) = (g * f^{-1})(p^{\alpha})$$

$$= g(p^{\alpha}) + g(p^{\alpha-1})f^{-1}(p) + g(p^{\alpha-2})f^{-1}(p^2)$$

$$= g(p^{\alpha-2}) \ [g(p)^2 + f^{-1}(p)g(p) + f^{-1}(p^2)] = 0 .$$

Thus, h is completely multiplicative by Proposition 1.9, and $f = g * h$. □

The functions of the preceding proposition can be characterized in several other ways.

Theorem 1.12. If f is a multiplicative function then the following statements are equivalent:

(1) f is a convolution of two completely multiplicative functions.

(2) There is a multiplicative function F such that for all m and n ,

$$f(mn) = \sum_{d|(m,n)} f(m/d)f(n/d)F(d) .$$

(3) There is a completely multiplicative function B such that for all m and n,

$$f(m)f(n) = \sum_{d|(m,n)} f(mn/d^2)B(d) .$$

(4) For all primes p and all $\alpha \geq 1$,

$$f(p^{\alpha+1}) = f(p)f(p^{\alpha}) + f(p^{\alpha-1})\,[f(p^2) - f(p)^2] .$$

Proof. In the course of the proof we will discover the relationship between the functions F and B, and show that they are uniquely determined by f.

(1) ⇒ (4) Suppose that $f = g * h$, where f and g are completely multiplicative. If $g(p) = M$ and $h(p) = N$ then $f(p) = M + N$ and $f(p^2) = M^2 + MN + N^2$. If $\alpha \geq 1$ then the right-hand side of the equation in (4) is

$$(M + N)\sum_{i=0}^{\alpha} M^i N^{\alpha-i} - MN\sum_{i=0}^{\alpha-1} M^i N^{\alpha-1-i}$$

$$= \sum_{i=0}^{\alpha} M^{i+1} N^{\alpha-i} + \sum_{i=0}^{\alpha} M^i N^{\alpha+1-i} - \sum_{i=0}^{\alpha-1} M^{i+1} N^{\alpha-i}$$

$$= M^{\alpha+1} + \sum_{i=0}^{\alpha} M^i N^{\alpha+1-i} = \sum_{i=0}^{\alpha+1} M^i N^{\alpha+1-i}$$

$$= f(p^{\alpha+1}) .$$

(4) ⇒ (1) For each prime p let M and N be the solutions of the quadratic equation

$$X^2 - f(p)X + f(p)^2 - f(p^2) = 0 .$$

M and N depend on p, of course, but since we work with one prime at a time there is no need for additional notation to indicate this dependence. Let g and h be the completely multiplicative functions such that for each p, $g(p) = M$ and $h(p) = N$. Then $f(p) = M + N = (g * h)(p)$ and for $\alpha \geq 2$,

$$\begin{aligned}(g * h)(p^\alpha) &= \sum_{i=0}^{\alpha} M^i N^{\alpha-i} \\ &= (M + N) \sum_{i=0}^{\alpha-1} M^i N^{\alpha-1-i} - MN \sum_{i=0}^{\alpha-2} M^i N^{\alpha-2-i} \\ &= f(p)f(p^{\alpha-1}) + f(p^{\alpha-2}) [f(p^2) - f(p)^2] \\ &= f(p^\alpha) .\end{aligned}$$

Therefore, since f is multiplicative, $f = g * h$.

(2) ⇒ (4) Let p be a prime and $\alpha \geq 1$. In the equation in (2) let $m = p^\alpha$ and $n = p$. The result is

$$f(p^{\alpha+1}) = f(p)f(p^\alpha) + f(p^{\alpha-1})F(p) .$$

For $\alpha = 1$, this yields $F(p) = f(p^2) - f(p)^2$.

(4) ⇒ (2) Assume that (4) holds. If $(mn, m'n') = 1$ then $((m,n),(m',n')) = 1$ and $(mm',nn') = (m,n)(m', n')$. Thus, to prove that the equation of (2) holds

for all m and n it is sufficient to show that there is a multiplicative function F such that for all primes p and all $\alpha, \beta \geq 1$,

$$f(p^{\alpha+\beta}) = \sum_{i=0}^{\min(\alpha,\beta)} f(p^{\alpha-i})f(p^{\beta-i})F(p^i) .$$

We shall show, in fact, that this is the case when $F = \mu B'$, where B' is the completely multiplicative function such that $B'(p) = f(p)^2 - f(p^2)$ for each prime p.

Without loss of generality we can assume that $\beta \leq \alpha$, and proceed by induction on β. The equation in (4) is the equation we want to prove when $\beta = 1$. Assume that $\beta > 1$ and that the equation holds when β is replaced by $\beta - 1$, for all $\alpha \geq \beta - 1$. Since $F = \mu B'$, $F(p^2) = F(p^3) = \ldots = 0$. Thus,

$$\begin{aligned}
f(p^{\alpha+\beta}) &= f(p^{\alpha+1+\beta-1}) \\
&= f(p^{\alpha+1})f(p^{\beta-1}) + f(p^{\alpha})f(p^{\beta-2})F(p) \\
&= [f(p)f(p^{\alpha}) - f(p^{\alpha-1})B'(p)]f(p^{\beta-1}) - f(p^{\alpha})f(p^{\beta-2})B'(p) \\
&= f(p^{\alpha}) [f(p)f(p^{\beta-1}) - f(p^{\beta-2})B'(p)] - f(p^{\alpha-1})f(p^{\beta-1})B'(p) \\
&= f(p^{\alpha})f(p^{\beta}) + f(p^{\alpha-1})f(p^{\beta-1})F(p) ,
\end{aligned}$$

which was to be proved.

Thus, (1), (2) and (4) have been proved to be equivalent. Furthermore, if they hold then for all m and n,

$$f(mn) = \sum_{d|(m,n)} f(m/d)f(n/d)\mu(d)B'(d) ,$$

where B' is defined as above.

(2) ⇒ (3) If (2) holds then for all m and n,

$$\sum_{d|(m,n)} f(mn/d^2)B'(d)$$

$$= \sum_{d|(m,n)} \sum_{D|(\frac{m}{d},\frac{n}{d})} f(\frac{m/d}{D})f(\frac{n/d}{D})\mu(D)B'(D)B'(d)$$

$$= \sum_{d|(m,n)} \sum_{\substack{e|(m,n)\\ d|e}} f(m/e)f(n/e)\mu(e/d)B'(e)$$

$$= \sum_{e|(m,n)} f(m/e)f(n/e)B'(e) \sum_{d|e} \mu(e/d)$$

$$= f(m)f(n) .$$

Thus, the completely multiplicative function B' serves as the completely multiplicative function B required in (3) .

(3) ⇒ (4) Suppose (3) holds. If p is a prime and if $m = n = p$, then we obtain

$$B(p) = f(p)^2 - f(p^2) .$$

Therefore, $B = B'$. Let $\alpha \geq 1$. If we take $m = p^{\alpha}$ and $n = p$, we obtain the equation in (4). □

In Exercise 1.64 the reader is asked to show that if (2) holds then it must necessarily be the case that $F = \mu B$. Recall that $\mu B = B^{-1}$. Sometimes we shall write B_f for B to indicate that B is associated with f. Note that it was shown in the course of the proof of Theorem 1.12 that if $f = g * h$, where g and h are completely multiplicative, then $B = gh$ (see the proof that (1) ⇒ (4) and observe that

$$\begin{aligned} B(p) &= f(p)^2 - f(p^2) \\ &= (M + N)^2 - (M^2 + MN + N^2) \\ &= MN = g(p)h(p)) . \end{aligned}$$

An arithmetical function that satisfies the conditions of Theorem 1.12 is called a <u>specially multiplicative function</u>. We have observed that for each positive integer k, the function σ_k is specially multiplicative. For this function, $B = \zeta_k$: if p is a prime then

$$\begin{aligned} B(p) &= \sigma_k(p)^2 - \sigma_k(p^2) \\ &= (1 + p^k)^2 - (1 + p^k + p^{2k}) \\ &= p^k = \zeta_k(p) . \end{aligned}$$

Therefore, for all m and n,

$$\sigma_k(mn) = \sum_{d|(m,n)} \sigma_k(m/d)\sigma_k(n/d)\mu(d)d^k$$

and

$$\sigma_k(m)\sigma_k(n) = \sum_{d|(m,n)} d^k \sigma_k(mn/d^2) .$$

The second of these identities was stated by E. Busche in 1906 and the first, with $k = 0$, by S. Ramanujan about ten years later. For this reason, either of the equivalent identities in (2) and (3) of Theorem 1.12 is called a Busche-Ramanujan identity.

There are other specially multiplicative functions, of course: we shall discuss some of them here and in the exercises.

The function β is defined by

$\beta(n)$ = the number of integers x such that $1 \leq x \leq n$ and (x,n) is a square.

If $1 \leq x \leq n$ and (x,n) is a square then for some square divisor d^2 of n , $x = d^2 y$ and $1 \leq y \leq n/d^2$ and $(y,n/d^2) = 1$. Thus,

$$\beta(n) = \sum_{d^2|n} \phi(n/d^2) \quad \text{for all } n .$$

Therefore, by Exercise 1.26, β is a multiplicative function. Furthermore, for a prime p and $\alpha \geq 1$,

$$\beta(p^\alpha) = \sum_{j \leq \frac{\alpha}{2}} \phi(p^{\alpha-2j})$$

$$= \begin{cases} p^\alpha - p^{\alpha-1} + p^{\alpha-2} - \ldots + p^2 - p + 1 & \text{if } \alpha \text{ is even} \\ p^\alpha - p^{\alpha-1} + p^{\alpha-2} - \ldots + p - 1 & \text{if } \alpha \text{ is odd} \end{cases}$$

$$= \sum_{j=0}^{\alpha} p^j \lambda(p^{\alpha-j}) ,$$

where λ is Liouville's function, defined in Exercise 1.47. Therefore,

$$\beta(n) = \sum_{d|n} d\lambda(n/d) \quad \text{for all} \quad n ,$$

$\beta = \zeta_1 * \lambda$. Since λ is completely multiplicative by Exercise 1.47 β is specially multiplicative. For this function, $B = \zeta_1\lambda$: it is the completely multiplicative function for which $B(p) = -p$ for all primes p . For all m and n ,

$$\beta(mn) = \sum_{d|(m,n)} d\lambda(d)\beta(m/d)\beta(n/d)$$

and

$$\beta(m)\beta(n) = \sum_{d|(m,n)} d\lambda(d)\beta(mn/d^2) .$$

Let $R(n)$ be the number of representations of n as a sum of two squares, i.e., the arithmetical function R is defined by

$R(n)$ = the number of ordered pairs of integers, positive, negative or zero, $\langle x,y \rangle$ such that $n = x^2 + y^2$.

For example, $R(5) = 8$ since

$$5 = 2^2 + 1^2 = 1^2 + 2^2 = (-2)^2 + 1^2 = 1^2 + (-2)^2$$

$$= 2^2 + (-1)^2 = (-1)^2 + 2^2 = (-2)^2 + (-1)^2 = (-1)^2 + (-2)^2 \; .$$

In some texts on the theory of numbers [1)] it is shown that for all n,

$$R_1(n) = \frac{1}{4} R(n) = \sum_{d|n} \chi(n) \; ,$$

where χ is the arithmetical function defined by

$$\chi(n) = \begin{cases} (-1)^{\frac{1}{2}(n-1)} & \text{if } n \text{ is odd} \\ 0 & \text{if } n \text{ is even} \; . \end{cases}$$

χ is completely multiplicative: it is called the <u>nonprincipal character (mod 4)</u>. Thus, the function R_1 is specially multiplicative, and $B = \chi$. If we write the Busche-Ramanujan identities for R_1, and multiply one by 4 and the other by 16, we find that for all m and n,

$$R(mn) = \frac{1}{4} \sum_{\substack{d|(m,n) \\ d \text{ odd}}} (-1)^{\frac{1}{2}(d-1)} R(m/d)R(n/d)\mu(d)$$

and

$$R(m)R(n) = 4 \sum_{\substack{d|(m,n) \\ d \text{ odd}}} (-1)^{\frac{1}{2}(d-1)} R(mn/d^2)$$

---

1) See [HW] pp. 241-243, or [NZ] pp. 106-110.

Exercises for Chapter 1

1.1. For all n ,

$$\sum_{j=1}^{n} \mu(j) \left[\frac{n}{j}\right] = 1$$

where, for a real number x , $[x]$ is the greatest integer not exceeding x .

1.2. The result of Exercise 1.1 can be generalized as follows. If f and g are arithmetical functions such that

$$f(n) = \sum_{d|n} g(d) \quad \text{for all } n ,$$

then

$$\sum_{j=1}^{n} g(j) \left[\frac{n}{j}\right] = \sum_{j=1}^{n} f(j) \quad \text{for all } n .$$

In particular, for all n ,

$$\sum_{j=1}^{n} \phi(j) \left[\frac{n}{j}\right] = \frac{1}{2} n(n + 1)$$

and

$$\sum_{j=1}^{n} \left[\frac{n}{j}\right] = \sum_{j=1}^{n} \tau(j) .$$

1.3. If f is an arithmetical function and if

$$g(n) = \sum_{r=1}^{n} f((n,r)) \quad \text{for all } n ,$$

then

$$\sum_{d|n} g(d) = \sum_{d|n} df(n/d) \quad \text{for all } n ,$$

i.e., $g * \zeta = f * \zeta_1$. Thus, $g = f * \phi$.

1.4. Let f and g be arithmetical functions and let h be the arithmetical function defined by

$$h(n) = \sum_{[a,b]=n} f(a)g(b) \quad \text{for all } n .$$

The sum is over all ordered pairs $\langle a,b \rangle$ of positive integers such that $[a,b] = n$. Then $h = (f * \zeta)(g * \zeta) * \mu$. This result has a natural extension to more than two functions.

1.5. For all n ,

$$\sum_{\substack{d|n \\ d \text{ squarefree}}} \mu(n/d) = \begin{cases} \mu(n^{\frac{1}{2}}) & \text{if } n \text{ is a square} \\ 0 & \text{otherwise.} \end{cases}$$

1.6. The <u>nth cyclotomic polynomial</u> $F_n(X)$ is the monic polynomial whose roots are the $\phi(n)$ complex primitive nth roots of unity, i.e.,

$$F_n(X) = \prod_{\substack{1\le h\le n \\ (h,n)=1}} (X - \eta^h) \ , \ \eta = e^{2\pi i/n}$$

For all n ,

$$X^n - 1 = \prod_{d|n} F_d(X)$$

and

$$F_n(X) = \prod_{d|n} (X^d - 1)^{\mu(n/d)} \quad .$$

1.7. For all n ,

$$\sum_{\substack{1\le h\le n \\ (h,n)=1}} (n, h - 1) = \phi(n)\tau(n) \ .$$

Note that the sum could be taken over all integers h in an arbitrary reduced residue system (mod n).

1.8. If f is a multiplicative function then for all n ,

$$\sum_{d|n} \mu(d)f(d) = \prod_{p|n} (1 - f(p))$$

(an empty product is equal to 1).

1.9. An arithmetical function f with $f(1) = 1$ is multiplicative if and only if for all m and n ,

$$f(m)f(n) = f((m,n))f([m,n]) \ .$$

1.10. An arithmetical function f with $f(1) \neq 0$ is multiplicative if and only if for all m and n,

$$f([m,n]/[d,e]) = f(m/d)f(n/e)$$

for all divisors d of m and e of n such that $(d,e) = (m,n)$.

1.11. Let f be a multiplicative function and suppose $f(k) \neq 0$. If g is defined by

$$g(n) = \frac{f(kn)}{f(k)} \quad \text{for all } n,$$

then g is a multiplicative function.

1.12. If g is a multiplicative function and if $f = g * \mu$, then

$$|\mu(n)|\ f(n) = \sum_{\substack{d|n \\ (d,n/d)=1}} g(d)|\mu(d)|\mu(n/d) \quad \text{for all } n.$$

In particular,

$$|\mu(n)|\phi(n) = \sum_{\substack{d|n \\ (d,n/d)=1}} d|\mu(d)|\mu(n/d) \quad \text{for all } n.$$

1.13. Let f be a multiplicative function. If $m|n$, $m \neq 1$, and $(m,n/m) = 1$, then

$$\sum_{d|m} f(d)f^{-1}(n/d) = 0.$$

1.14. The <u>core function</u> γ is defined by

$$\gamma(n) = \begin{cases} 1 & \text{if } n = 1 \\ p_1 \cdots p_t & \text{if } n = p_1^{\alpha_1} \cdots p_t^{\alpha_t} . \end{cases}$$

γ is a multiplicative function and

$$\gamma(n) = \sum_{d|n} |\mu(d)|\phi(d) \quad \text{for all } n .$$

1.15. If f and g are arithmetical functions then

$$f(n) = \sum_{\substack{d|n \\ \gamma(d)=\gamma(n)}} g(d) \quad \text{for all } n$$

if and only if

$$g(n) = \sum_{\substack{d|n \\ \gamma(d)=\gamma(n)}} f(d)\mu(n/d) \quad \text{for all } n .$$

1.16. If f is a multiplicative function then for all n, r and s such that $\gamma(n) \nmid rs$,

$$\sum_{d|n} f(rd)f^{-1}(sn/d) = 0 .$$

1.17. If f is a multiplicative function then for all n,

$$\sum_{\substack{d|n \\ \gamma(d)=\gamma(n)}} f(n/d)f^{-1}(d) = (-1)^{\omega(n)} f(n) ,$$

where $\omega(n)$ is the number of distinct primes that divide n (so that $\omega(1) = 0$).

1.18. If f is a multiplicative function then for all n and r,

$$\sum_{d|n} f(rn/d)f^{-1}(d) = (-1)^{\omega(n)} \sum_{\substack{e|r \\ \gamma(e)=\gamma(n)}} f(r/e)f^{-1}(ne).$$

1.19. If A is the $n \times n$ matrix $[(i,j)]$, $1 \leq i,j \leq n$, having (i,j) in the ith row and jth column, then $\det A = \phi(1)\phi(2)\ldots\phi(n)$. This determinant is called <u>Smith's determinant</u>.

1.20. Let k be a positive integer. The arithmetical function δ_k is defined by

$$\delta_k(n) = \text{the greatest divisor } d \text{ of } n \text{ such that } (d,k) = 1 .$$

δ_k is a multiplicative function and

$$\delta_k(n) = \sum_{\substack{d|n \\ (d,k)=1}} \phi(d) \quad \text{for all } n .$$

1.21. Let $f(X)$ be a polynomial with integer coefficients. The arithmetical function ϕ_f is defined by

$$\phi_f(n) = \text{the number of integers } x \text{ such that } 1 \leq x \leq n \text{ and } (f(x),n) = 1 .$$

Thus, if $f(X) = X$ then $\phi_f = \phi$. The function ϕ_f is multiplicative. For all n,

$$\phi_f(n) = n \prod_{p|n} \left(1 - \frac{f_p}{p}\right)$$

where, for a prime p, f_p is the number of solutions of the congruence $f(X) \equiv 0 \pmod{p}$. Let ν_f be the multiplicative function such that for all primes p and all $\alpha \geq 1$, $\nu_f(p^\alpha) = f_p^\alpha$. For all n,

$$\sum_{d|n} \phi_f(d)\nu_f(n/d) = n$$

and

$$\phi_f(n) = \sum_{d|n} d\nu_f(n/d)\mu(n/d) .$$

1.22. Let $m_1, \ldots, m_t$ be integers. For each prime p let $N(p)$ be the number of distinct residue classes (mod p) represented by these integers. Then the number of sequences $x + m_1, \ldots, x + m_t$ such that $1 \leq x \leq n$ and $(x + m_i, n) = 1$ for $i = 1, \ldots, t$, is equal to

$$n \prod_{p|n} (1 - \frac{N(p)}{p}) .$$

(Hint: make a judicious choice of $f(X)$ in Exercise 1.21.)

1.23. Let k be a nonnegative integer and define the arithmetical function $\phi(\cdot, k)$ by

$$\phi(n, k) = \text{the number of integers } x \text{ such that } 1 \leq x \leq n \text{ and } (x, n) = (n + k - x, n) = 1.$$

$\phi(\cdot, 0) = \phi$, and the function $\phi(\cdot, 1)$ is called <u>Schemmel's function</u>. If $(m,n) = 1$ then $\phi(mn, k) = \phi(m, k)\phi(n, k)$, i.e., $\phi(\cdot, k)$ is a multiplicative function. For all n ,

$$\phi(n, k) = n \prod_{p|n} (1 - \frac{\varepsilon(p)}{p}) ,$$

where

$$\varepsilon(p) = \begin{cases} 1 & \text{if } p \mid k \\ 2 & \text{if } p \nmid k . \end{cases}$$

1.24. The arithmetical function θ is defined by

$\theta(n)$ = the number of ordered pairs $\langle a, b \rangle$ of positive integers such that $(a, b) = 1$ and $n = ab$.

θ is a multiplicative function and for all n ,

$$\theta(n) = 2^{\omega(n)} .$$

For all n , $\theta(n)$ is the number of squarefree divisors of n : hence

$$\theta(n) = \sum_{d \mid n} |\mu(d)| \quad \text{for all } n .$$

1.25. Let k be a positive integer. If f and g are arithmetical functions then

$$f(n) = \sum_{d^k \mid n} g(n/d^k) \quad \text{for all } n$$

if and only if

$$f(n) = \sum_{d^k|n} \mu(d)f(n/d^k) \quad \text{for all} \quad n .$$

1.26. Let k be a positive integer. If g and h are multiplicative functions then the arithmetical function f defined by

$$f(n) = \sum_{d^k|n} g(d)h(n/d^k) \quad \text{for all} \quad n$$

is multiplicative.

1.27. Let k be an integer, $k \geq 2$. The arithmetical function θ_k is defined by

$$\theta_k(n) = \text{the number of k-free divisors of } n .$$

Thus $\theta_2 = \theta$. The function θ_k is multiplicative and for all n ,

$$\theta_{2k}(n) = \text{the number of ordered pairs } \langle a, b \rangle \text{ of positive integers such that } (a, b)_k = 1 \text{ and } n = ab ,$$

where $(a, b)_k$ is the largest common kth power divisor of a and b . For all n ,

$$\sum_{d^k|n} \theta_k(n/d^k) = \tau(n)$$

and

$$\theta_k(n) = \sum_{d^k|n} \mu(d)\tau(n/d^k) .$$

1.28. Let x be a positive real number and for all n let

$\phi(x, n)$ = the number of integers y such that $1 \le y \le x$ and $(y, n) = 1$.

Thus $\phi(n, n) = \phi(n)$. For all x and n ,

$$\phi(x, n) = \sum_{d|n} \mu(d) \left[\frac{x}{d}\right]$$

and

$$\left|\phi(x, n) - \frac{x\phi(n)}{n}\right| \le \theta(n) .$$

1.29. Let k be a positive integer. <u>Klee's function</u> Φ_k is defined by

$\Phi_k(n)$ = the number of integers x such that $1 \le x \le n$ and $(x, n)_k = 1$.

($(a, b)_k$ is defined in Exercise 1.27 for $k \ge 2$, and $(a, b)_1 = (a, b)$.) Thus $\Phi_1 = \phi$. For all n ,

$$\sum_{d^k|n} \Phi_k(n/d^k) = n$$

and

$$\Phi_k(n) = n \sum_{d^k | n} \mu(d)/d^k .$$

(Hint: formulate and prove a lemma analogous to Lemma 1.4.) Φ_k is a multiplicative function and for all n,

$$\Phi_k(n) = n \prod_{p^k | n} \left(1 - \frac{1}{p^k}\right) .$$

1.30. Let k be a positive integer. The arithmetical function μ_k is defined by

$$\mu_k(n) = \begin{cases} 1 & \text{if } n = 1 \\ (-1)^t & \text{if } n = p_1{}^k \cdots p_t{}^k \\ 0 & \text{otherwise} . \end{cases}$$

Thus $\mu_1 = \mu$. The function μ_k is multiplicative and

$$\sum_{d|n} \mu_k(d) = \begin{cases} 1 & \text{if } n \text{ is k-free} \\ 0 & \text{otherwise} . \end{cases}$$

Furthermore, $\Phi_k = \zeta_1 * \mu_k$.

1.31. Let k be a positive integer. For all n,

$$\Phi_k(n) = \sum_{\substack{d|n \\ d \text{ k-free}}} \phi(n/d)$$

and

$$\sum_{d|n} \Phi_k(d) = \sum_{d^k|n} \sigma(n/d^k)\mu(d)$$

$$= n \sum_{\substack{d|n \\ d \text{ k-free}}} \frac{1}{d} .$$

1.32. Let k be an integer, $k \geq 2$. The arithmetical function τ_k is defined by

$\tau_k(n)$ = the number of ordered k-tuples $\langle a_1, \ldots, a_k \rangle$ of positive integers such that $n = a_1 \ldots a_k$.

Thus $\tau_2 = \tau$. The function τ_k is multiplicative and for every prime p and all $\alpha \geq 1$,

$$\tau_k(p^\alpha) = \binom{\alpha + k - 1}{\alpha} .$$

For all n , and $k \geq 3$,

$$\tau_k(n) = \sum_{d|n} \tau_{k-1}(d) .$$

1.33. Let h and k be positive integers, $k \geq 2$. The arithmetical function $\tau_{k,h}$ is defined by

$\tau_{k,h}(n)$ = the number of ordered k-tuples $\langle a_1, \ldots, a_k \rangle$ of positive integers such that each a_i is an hth power and $n = a_1 \ldots a_k$.

For all n ,

$$\tau_{k,h}(n) = \begin{cases} \tau_k(n^{\frac{1}{h}}) & \text{if } n \text{ is an } h\text{th power} \\ 0 & \text{otherwise .} \end{cases}$$

$\tau_{k,h}$ is a multiplicative function and for all n , and $k \geq 3$,

$$\sum_{d|n} \tau_{k,h}(d)\Phi_h(n/d) = \sum_{d|n} d\tau_{k-1,h}(n/d) ,$$

where Φ_h is Klee's function .

1.34. Let k be a positive integer. The arithmetical function ψ_k is defined by

$$\psi_k(n) = \sum_{d|n} d^k|\mu(n/d)| \quad \text{for all } n .$$

$\psi = \psi_1$ is called <u>Dedekind's function</u>. ψ_k is a multiplicative function and for all n ,

$$\psi_k(n) = \sum_{d|n} \left(\frac{d}{(d,n/d)}\right)^k J_k((d,n/d))$$

$$= n^k \prod_{p|n} (1 + \frac{1}{p^k}) = J_{2k}(n)/J_k(n) .$$

1.35. Let k be a positive integer. The arithmetical function q_k is defined by

$$q_k(n) = \begin{cases} |\mu(n^{\frac{1}{k}})| & \text{if } n \text{ is a } k\text{th power} \\ 0 & \text{otherwise.} \end{cases}$$

q_k is a multiplicative function and $q_1(n) = 1$ or 0 according as n is or is not squarefree. If $\Psi_k = \zeta_1 * q_k$ then $\Psi_1 = \psi$ and for all n,

$$\Psi_k(n) = n \prod_{p^k | n} \left(1 + \frac{1}{p^k}\right) .$$

1.36. Let k be a positive integer. A complete set of residues (mod n^k) is called an <u>(n,k)-residue system</u>: $\{x : 1 \le x \le n^k\}$ is the <u>minimal</u> (n,k)-residue system. The set of all x in an (n,k)-residue system such that $(x,n^k)_k = 1$ is called a <u>reduced (n,k)-residue system</u>: $\{x : 1 \le x \le n^k$ and $(x,n^k)_k = 1\}$ is the <u>minimal</u> reduced (n,k)-residue system. If $d|n$ let

$$S_d = \{x(n/d)^k : x \text{ belongs to the minimal reduced } (d,k)\text{-residue system}\}.$$

If $d,e|n$ and $d \neq e$ then $S_d \cap S_e$ is empty, and

$$\bigcup_{d|n} S_d = \{1, \ldots, n^k\} .$$

1.37. Let k be a positive integer. The arithmetical function ϕ_k is defined by

$$\phi_k(n) = \text{the number of integers in a reduced } (n,k)\text{-residue system} .$$

Thus $\phi_1 = \phi$. For all n,

$$\sum_{d|n} \phi_k(d) = n^k$$

and

$$\phi_k(n) = \sum_{d|n} d^k \mu(n/d) ,$$

i.e., $\phi_k = \zeta_k * \mu = J_k = H_k$.

1.38. Let k be a positive integer. If $(m,n) = 1$ then the set of integers $xm^k + yn^k$, where x runs over a reduced (n,k)-residue system and y runs over a reduced (m,k)-residue system, is a reduced (mn,k)-residue system.

1.39. Let k be a positive integer. If $d|n$ then every reduced (n,k)-residue system is the union of $\phi_k(n)/\phi_k(d)$ pairwise disjoint reduced (d,k)-residue systems. In particular, every reduced residue system (mod n) is the union of $\phi(n)/\phi(d)$ pairwise disjoint reduced residue systems (mod d). (Hint: consider the three cases $(d,n/d) = 1$, $\gamma(d)|\gamma(n/d)$ and the general case.)

1.40. Let k be a positive integer. If $d|n^k$ and $(d,n^k)_k = 1$ let

$$R_d = \{xd : 1 \leq x \leq n^k/d \text{ and } (x, n^k/d) = 1\} .$$

If $d, e|n^k$, $(d,n^k)_k = (e,n^k)_k = 1$ and $d \neq e$, then $R_d \cap R_e$ is empty, and

$$\bigcup_{\substack{d|n^k \\ (d,n^k)_k=1}} R_d = \{x : 1 \leq x \leq n^k \text{ and } (x,n^k)_k = 1\} .$$

Thus, for all n,

$$\phi_k(n) = \sum_{\substack{d \mid n^k \\ (d,n^k)_k=1}} \phi(n^k/d) .$$

1.41. Let k be a positive integer. The set of ordered k-tuples $\langle x_1, \ldots, x_k \rangle$ obtained when each x_i runs over a complete set of residues (mod n) is called a k-residue system (mod n). The set of k-tuples in a k-residue system (mod n) such that $(x_1, \ldots, x_k, n) = 1$ is called a reduced k-residue system (mod n). If $d \mid n$ then every reduced k-residue system (mod n) is the union of $J_k(n)/J_k(d)$ pairwise disjoint reduced k-residue systems (mod d).

1.42. Proposition 1.7 can be proved by using the result in Exercise 1.4.

1.43. Let k be a positive integer and suppose that $k = k_1 + \ldots + k_t$, where $k_i \geq 1$ for $i = 1, \ldots, t$. For all n,

$$J_k(n) = \sum_{[e_1, \ldots, e_t]=n} J_{k_1}(e_1) \ldots J_{k_t}(e_t) .$$

1.44. If f is a multiplicative function and if $f * f = f\tau$, then f is completely multiplicative.

1.45. Let f be a multiplicative function. If there is a completely multiplicative function g, with $g(p)^\alpha \neq 1$ for all primes p and all $\alpha \geq 1$, such that $f(g * \mu) = fg * f^{-1}$, then f is completely multiplicative. For example, if $f\phi = f\zeta_1 * f^{-1}$ then f is completely multiplicative.

1.46. A multiplicative function f is completely multiplicative if and only if $(fg)^{-1} = fg^{-1}$ for all arithmetical functions g that have inverses.

1.47. <u>Liouville's function</u> λ is defined by $\lambda(1) = 1$ and

$$\lambda(n) = (-1)^{\alpha_1+\dots+\alpha_t} \quad \text{if} \quad n = p_1^{\alpha_1} \dots p_t^{\alpha_t} .$$

λ is completely multiplicative and for all n,

$$\sum_{d|n} \lambda(d) = \begin{cases} 1 & \text{if } n \text{ is a square} \\ 0 & \text{otherwise} . \end{cases}$$

Thus, for all n,

$$\lambda(n) = \sum_{d^2|n} \mu(n/d^2) .$$

Furthermore, for all n,

$$\sum_{d^2|n} \lambda(n)\tau(n/d) = \text{the number of square divisors of } n .$$

Finally, $\theta^{-1} = \lambda\theta$ (see Exercise 1.24).

1.48. In this exercise, and in the ones that follow through Exercise 1.59, the problem is to verify the stated identity. This can be done by using the fact that the functions involved are multiplicative, and many of them can also be verified by using properties of the convolution of arithmetical functions. In these exercises, k, s and t are nonnegative integers: the index on a Jordan function must be positive. For all n,

$$\sum_{d|n} d^s\sigma_k(d) = \sum_{d|n} d^{s+k}\sigma_s(n/d)$$

$$= \sum_{d|n} d^s\sigma_{s+k}(n/d) .$$

1.49. For all n ,

$$\sum_{d|n} d^s\sigma_{k+t}(d)\sigma_k(n/d) = \sum_{d|n} d^k\sigma_{s+t}(d)\sigma_s(n/d) .$$

1.50. For all n ,

$$\sum_{d|n} d^s J_k(d)\sigma_s(n/d) = \sigma_{k+s}(n) .$$

1.51. For all n ,

$$\sum_{d|n} J_k(d)\sigma_s(n/d) = \begin{cases} n^s\sigma_{k-s}(n) & \text{if } k \geq s \\ n^k\sigma_{s-k}(n) & \text{if } k \leq s . \end{cases}$$

1.52. For all n ,

$$\sum_{d|n} \theta(d) = \tau(n^2) , \quad \sum_{d^2|n} \theta(n/d^2) = \tau(n)$$

and

$$\theta(n) = \sum_{d|n} \tau(d^2)\mu(n/d) .$$

1.53. For all n,

$$\sum_{d|n} \theta(d)^k \sigma_s(n/d) = \sum_{d|n} d^s \tau((n/d)^{2^k}) .$$

1.54. For all n,

$$\sum_{d|n} \theta(d)^k = \tau(n^{2^k}) .$$

1.55. For all n,

$$\sum_{d|n} \tau(d^{2^k})\theta(n/d) = \sum_{d|n} \tau(d^2)\theta(n/d)^k .$$

1.56. For all n,

$$\sum_{d|n} \lambda(d)\tau(d^2)\sigma_k(n/d) = \sum_{d|n} d^k \tau(n/d)\lambda(n/d) .$$

1.57. For all n,

$$\sum_{d|n} \tau(d^{2^k})\lambda(n/d) = \sum_{d^2|n} \theta(n/d^2)^k .$$

1.58. For all n,

$$\sum_{d^2|n} \lambda(d)\tau(n/d^2) = \sum_{d^4|n} \theta(n/d^4) .$$

1.59. For all n,

$$\sum_{d|n} \lambda(d)\sigma_k(n/d) = \sum_{d^2|n} (n/d^2)^k .$$

1.60. Let f and g be arithmetical functions. If

$$F(n) = \sum_{d|n} f(d) \quad \text{and} \quad G(n) = \sum_{d|n} g(d) \quad \text{for all } n,$$

then

$$\sum_{d|n} f(d)G(n/d) = \sum_{d|n} g(d)F(n/d) \quad \text{for all } n.$$

1.61. Let k be a positive integer. A multiplicative function f is the convolution of k completely multiplicative functions if and only if $f^{-1}(p^\alpha) = 0$ for all primes p and all $\alpha \geq k + 1$.

1.62. Let $g_1,\ldots,g_k$ be completely multiplicative functions, and $f = g_1*\ldots*g_k$. If n is the product of distinct primes then $f^{-1}(n^k) = \mu(n)^k g_1(n)\ldots g_k(n)$.

1.63. If g_1, g_2, h_1 and h_2 are completely multiplicative functions then

$$g_1g_2 * g_1h_2 * h_1g_2 * h_1h_2 = (g_1 * h_1)(g_2 * h_2) * u$$

where, for all n,

$$u(n) = \begin{cases} g_1h_1g_2h_2(n^{\frac{1}{2}}) & \text{if } n \text{ is a square} \\ 0 & \text{otherwise.} \end{cases}$$

1.64. If (2) of Theorem 1.12 holds for a multiplicative function f then $F = \mu B$.

1.65. If f is a specially multiplicative function and h is a completely multiplicative function, then hf is specially multiplicative and $B_{hf} = h^2 B_f$, where $h^2 = hh$.

1.66. A specially multiplicative function f is completely multiplicative if and only if $B_f = \pm\,\delta$.

1.67. Let V be the function of two positive integer variables defined by

$$V(m,n) = \begin{cases} (-1)^{\omega(n)} & \text{if } \gamma(m) = \gamma(n) \\ 0 & \text{otherwise.} \end{cases}$$

If f is a multiplicative function then for all m and n,

$$f(mn) = \sum_{d|m} \sum_{e|n} f(m/d)f(n/e)f^{-1}(de)V(d,e)$$

(Hint: use Exercise 1.18.)

1.68. The identity in Exercise 1.67 can be proved directly by showing that it holds whenever m and n are powers of the same prime, and by observing that it is sufficient to consider only this special case.

1.69. If f is a specially multiplicative function then the identity of Exercise 1.67 is the identity of Theorem 1.12(2). Thus, we have an alternate proof that (1) implies (2) in that theorem.

1.70. The <u>norm</u> of a multiplicative function f is the arithmetical function $N(f)$ defined by

$$N(f)(n) = \sum_{d|n^2} f(n^2/d)\lambda(d)f(d) \quad \text{for all } n.$$

$N(f)$ is multiplicative, and if f is completely multiplicative, so is $N(f)$. In fact, if f is completely multiplicative then $N(f) = f^2$.

1.71. If f and g are multiplicative functions then $N(f * g) = N(f) * N(g)$.

1.72. If f is a specially multiplicative function, so is $N(f)$, and $B_{N(f)} = B_f^2$.

1.73. If f is a multiplicative function and $f' = f * \lambda f$, then

$$f'(n) = \begin{cases} N(f)(n^{\frac{1}{2}}) & \text{if } n \text{ is a square} \\ 0 & \text{otherwise.} \end{cases}$$

1.74. If f is specially multiplicative function then $f^2 = N(f) * \theta B = N(f) * \mu^2 B * B$.

1.75. If f is a specially multiplicative function then $f^2 * \lambda f^2 = N(f) * \lambda N(f)$.

1.76. What are the arithmetical functions $N(\zeta_k)$, $N(\lambda)$, $N(\sigma_k)$, $N(\beta)$, $N(\chi)$ and $N(R_1)$?

1.77. What are the arithmetical functions $N(\mu)$ and $N(\phi)$?

1.78. Let k be a positive integer. The arithmetical function β_k is defined by

$\beta_k(n)$ = the number of integers x such that $1 \le x \le n^k$ and $(x, n^k)_k$ is a 2kth power.

For all n ,

$$\beta_k(n) = \sum_{d^2|n} \phi_k(n/d^2) = \sum_{d|n} d^k \lambda(n/d)$$

Thus, β_k is specially multiplicative, and $B = \zeta_k \lambda$. What is $N(\beta_k)$? What are the Busche-Ramanujan identities for β_k?

1.79. Let f be a specially multiplicative function. Let g be an arithmetical function and let $G = g * \mu$. For all m and n ,

$$\sum_{d|(m,n)} G(d)B(d)f(m/d)f(n/d)$$

$$= \sum_{d|(m,n)} g(d)B(d)f(mn/d^2) .$$

In particular, for a positive integer k ,

$$\sum_{d|(m,n)} J_k(d)B(d)f(m/d)f(n/d)$$

$$= \sum_{d|(m,n)} d^k B(d)f(mn/d^2) ,$$

and more particularly, for a nonnegative integer h,

$$\sum_{d|(m,n)} d^h J_k(d)\sigma_h(m/d)\sigma_h(n/d)$$

$$= \sum_{d|(m,n)} d^{h+k}\sigma_h(mn/d^2).$$

1.80. A divisor d of n is called a <u>block factor</u> of n if $(d,n/d) = 1$. (In Chapter 4, such a divisor of n is called a <u>unitary divisor</u>, but we use the term block factor here for historical reasons: see the notes following the exercises.) Let k be a positive integer. If m and n have no common block factor (other than 1, of course) then

$$J_k(mn) = \sum_{d|(m,n)} d^k J_k(m/d)J_k(n/d).$$

This can be proved directly, using the fact that J_k is a multiplicative function.

1.81. If, in the identity of Exercise 1.67, $f = J_k$ and m and n have no common block factor, the result is the identity of Exercise 1.80.

1.82. Let f be an arithmetical function. A <u>restricted Busche-Ramanujan identity</u> holds for f if there is a multiplicative function F such that if m and n have no common block factor then

$$f(mn) = \sum_{d|(m,n)} f(m/d)f(n/d)F(d).$$

A function f is called a <u>totient</u> if there exist completely multiplicative functions g and h such that $f = g * h^{-1}$. If f is a totient then a restricted Busche-Ramanujan identity holds for f, with $F = gh$.

1.83. The <u>kth convolute</u> $\Omega_k(f)$ of an arithmetical function f is defined by

$$\Omega_k(f)(n) = \begin{cases} f(n^{\frac{1}{k}}) & \text{if } n \text{ is a kth power} \\ 0 & \text{otherwise.} \end{cases}$$

For example, if μ_k is the function defined in Exercise 1.30 then $\mu_k = \Omega_k(\mu)$. Convolutes of other particular functions occur in Exercises 1.5, 1.33, 1.35, 1.47, 1.63, 1.73, 1.89 and 1.91. If f and g are arithmetical functions then $\Omega_k(f + g) = \Omega_k(f) + \Omega_k(g)$, $\Omega_k(fg) = \Omega_k(f)\Omega_k(g)$, $\Omega_k(f * g) = \Omega_k(f) * \Omega_k(g)$, and if f has an inverse, so does $\Omega_k(f)$, and $\Omega_k(f^{-1}) = \Omega_k(f)^{-1}$. If f is a totient then a restricted Busche-Ramanujan identity holds for $\Omega_2(f)$.

1.84. A multiplicative function f is a <u>cross between</u> functions in a set $\{f_i : i \in I\}$ of arithmetical functions if for each prime p there is an $i \in I$ such that $f(p^\alpha) = f_i(p^\alpha)$ for all $\alpha \geq 1$. If f is a cross between specially multiplicative functions, totients and the 2nd convolutes of totients, then a restricted Busche-Ramanujan identity holds for f.

1.85. Let f be a multiplicative function and assume that a restricted Busche-Ramanujan identity holds for f. Then f is a cross between specially multiplicative functions, totients and the second convolutes of totients.

1.86. Let f and h be arithmetical functions and assume that a restricted Busche-Ramanujan identity holds for f. Let $H = h * \zeta$. If m and n have no common block factor then

$$\sum_{d|(m,n)} H(d)F(d)f(m/d)f(m/d) = \sum_{d|(m,n)} h(d)F(d)f(mn/d^2).$$

1.87. Assume that a restricted Busche-Ramanujan holds for the arithmetical function f. If m and n have no common block factor then

$$f(m)f(n) = \sum_{d|(m,n)} \mu(d)F(d)f(mn/d^2).$$

(Hint: let $h = \mu$ in Exercise 1.86.) In particular, if k is a positive integer and if m and n have no common block factor, then

$$J_k(m)J_k(n) = \sum_{d|(m,n)} \mu(d)d^k J_k(mn/d^2).$$

1.88. Let h be a completely multiplicative function and let $f = h * \zeta$ and $g = h * \mu$. For all n,

$$\sum_{d|n} h(n/d)g(n/d)f(d)f(nd) = \sum_{d|n} h^2(n/d)f(nd^2)$$

and

$$\sum_{d|n} h(n/d)f(n/d)g(d)g(nd) = \sum_{d|n} h^2(n/d)g(nd^2).$$

(Hint: in Exercises 1.79 and 1.86 let $m = n^2$.) In particular, if k is a positive integer then for all n,

$$\sum_{d|n} \frac{J_k(n/d)}{d^k} \sigma_k(d)\sigma_k(nd) = n^k \sum_{d|n} \frac{\sigma_k(nd^2)}{d^{2k}}$$

and

$$\sum_{d|n} \frac{\sigma_k(n/d)}{d^k} J_k(d)J_k(nd) = n^k \sum_{d|n} \frac{J_k(nd^2)}{d^{2k}} .$$

1.89. Let k be a nonnegative integer and s a positive integer. Gegenbauer's function $\rho_{k,s}$ is defined by

$$\rho_{k,s}(n) = \sum_{\substack{d|n \\ \frac{n}{d} \text{ is an sth power}}} d^k \quad \text{for all } n.$$

Thus, $\rho_{k,s} = \zeta_k * \nu_s$ where

$$\nu_s(n) = \begin{cases} 1 & \text{if } n \text{ is an sth power} \\ 0 & \text{otherwise.} \end{cases}$$

ν_s is the sth convolute of ζ. For all n,

$$\sum_{d|n} d^h \rho_{k,s}(n/d) = \sum_{d|n} d^k \rho_{h,s}(n/d)$$

and

$$\sum_{d|n} d^k J_h(d)\rho_{k,s}(n/d) = \rho_{h+k,s}(n).$$

1.90. If h, k and s are positive integers with $h \geq k$ then for all n

$$\sum_{d^s|n} J_{sk}(d)\rho_{h,s}(n/d^s) = n^k\rho_{h-k,s}(n)$$

and

$$\sum_{d|n} \rho_{h,s}(d)\rho_{k,s}(n/d) = n^k \sum_{d^s|n} \tau(d)\sigma_{h-k}(n/d^s)/d^{ks}.$$

1.91. For all n,

$$\sum_{d|n} d^k\lambda(d)\rho_{k,2s}(n/d) = \sum_{d|n} \lambda(d)\rho_{k,s}(d)\rho_{k,s}(n/d)$$

$$= \begin{cases} \rho_{2k,2}(n^{\frac{1}{2}}) & \text{if } n \text{ is a square} \\ 0 & \text{otherwise} \end{cases}$$

and

$$\sum_{d^s|n} \lambda(d)\rho_{k,s}(n/d^s) = \rho_{k,2s}(n).$$

1.92. Let q and k be integers with $0 < q < k$, and let $S_{k,q}$ be the set of all positive integers such that if $n = p_1^{\alpha_1} \cdots p_t^{\alpha_t}$ then, for $i = 1,\ldots,t$, $\alpha_i \equiv 0,1,\ldots,$ or $q-1 \pmod k$. Then $n \in S_{k,q}$ if and only if whenever $n = m^k r$, where r is k-free, r is q-free. The integer r is called the <u>k-free part</u> of n. Thus, $n \in S_{k,q}$ if and only if its k-free part is q-free. The integers in $S_{k,q}$ are called <u>**(k,q)-integers**</u>. Let $\lambda_{k,q}$ be the multiplicative function such that for all primes p and all $\alpha \geq 1$,

$$\lambda_{k,q}(p^{\alpha}) = \begin{cases} 1 & \text{if } \alpha \equiv 0 \pmod{k} \\ -1 & \text{if } \alpha \equiv q \pmod{k} \\ 0 & \text{otherwise} . \end{cases}$$

Then $\zeta * \lambda_{k,q} = \zeta_{k,q}$, where

$$\zeta_{k,q}(n) = \begin{cases} 1 & \text{if } n \in S_{k,q} \\ 0 & \text{if } n \notin S_{k,q} \end{cases}$$

Note that $\lambda_{2,1} = \lambda$, Liouville's function.

1.93. Continuing Exercise 1.92, let $\mu_{k,q}$ be the multiplicative function such that for all primes p and all $\alpha \geq 1$,

$$\text{if } q|k \text{ then } \mu_{k,q}(p^{\alpha}) = \begin{cases} 1 & \text{if } 0 \leq \alpha \leq k - q \\ 0 & \text{otherwise} , \end{cases}$$

$$\text{if } q \nmid k \text{ then } \mu_{k,q}(p^{\alpha}) = \begin{cases} 1 & \text{if } \alpha \equiv 0 \pmod{q} \\ -1 & \text{if } \alpha \geq k \text{ and } \alpha \equiv k \pmod{q} \\ 0 & \text{otherwise} . \end{cases}$$

Then $\lambda_{k,q}^{-1} = \mu_{k,q}$.

1.94. Continuing Exercise 1.93, if f is an arithmetical function and $F = \zeta * f$, then

$$g(n) = \sum_{\substack{d|n \\ d \in S_{k,q}}} f(n/d) \quad \text{for all } n$$

if and only if

$$g(n) = \sum_{d|n} \lambda_{k,q}(d) F(n/d) \quad \text{for all } n .$$

1.95. Continuing Exercise 1.94, let $\phi_{k,q}$ be the arithmetical function defined by

$$\phi_{k,q}(n) = \text{the number of integers } x \text{ such that } 1 \le x \le n \text{ and } (x,n) \in S_{k,q} .$$

For all n,

$$\phi_{k,q}(n) = n \sum_{d|n} \lambda_{k,q}(d)/d = \sum_{\substack{d|n \\ d \in S_{k,q}}} \phi(n/d) .$$

1.96. Continuing Exercise 1.95, let $\theta_{k,q}$ be the arithmetical function defined by

$$\theta_{k,q}(n) = \text{the number of divisors } d \text{ of } n \text{ with } d \in S_{k,q}.$$

For all n,

$$\theta_{k,q}(n) = \sum_{d^k|n} \theta_q(n/d^k).$$

1.97. With a sequence $\{a_n\}$, $n = 0,1,2,\ldots,$ of complex numbers we can associate a <u>formal power series</u>

$$a(X) = \sum_{n=0}^{\infty} a_n X^n.$$

Equality of formal power series means term-by-term equality. The <u>sum</u> and <u>product</u> of $a(X)$ and

$$b(X) = \sum_{n=0}^{\infty} b_n X^n$$

are defined to be

$$a(X) + b(X) = \sum_{n=0}^{\infty} (a_n + b_n) X^n$$

and

$$a(X)b(X) = \sum_{n=0}^{\infty} c_n X^n, \quad \text{where} \quad c_n = \sum_{k=0}^{n} a_k b_{n-k}.$$

The set of formal power series, together with these two binary operations, is a commutative ring with unity, called the <u>ring of formal power series</u> (over the field of complex numbers). If a is a complex number then the formal power series $a(X) = 1 - aX$ is a unit in this ring, i.e., it has an inverse with respect to multiplication. In fact, its inverse is the formal power series

$$\sum_{n=0}^{\infty} a^n X^n, \quad \text{where} \quad a^0 = X^0 = 1.$$

Thus, in the ring of formal power series,

$$\sum_{n=0}^{\infty} a^n X^n = \frac{1}{1 - aX}.$$

1.98. Let f be an arithmetical function and let p be a prime. The formal power series

$$f_p(X) = \sum_{n=0}^{\infty} f(p^n)X^n$$

is called the <u>Bell series of</u> f <u>relative to</u> p. If f and g are multiplicative functions then $f = g$ if and only if $f_p(X) = g_p(X)$ for all primes p. For each prime p, $\mu_p(X) = 1 - X$, and if k is a positive integer then

$$(J_k)_p(X) = \frac{1 - X}{1 - p^k X} .$$

1.99. For all primes p,

$$(\zeta_k)_p = \frac{1}{1 - p^k X}, \qquad (\mu^2)_p(X) = 1 + X,$$

$$\lambda_p(X) = \frac{1}{1 + X}, \qquad (\sigma_k)_p(X) = \frac{1}{1 - (p^k + 1)X + p^k X^2},$$

$$\theta_p(X) = \frac{1 + X}{1 - X}, \qquad (\beta_k)_p(X) = \frac{1}{1 - (p^k - 1)X - p^k X^2},$$

$$(\rho_{k,s})_p(X) = \frac{1}{1 - p^k X - X^s + p^k X^{s+1}} .$$

1.100. If f and g are arithmetical functions then for all primes p, $(f * g)_p(X) = f_p(X)g_p(X)$. Many of the identities obtained in this chapter follow immediately from this result and the formulas in Exercise 1.99 and similar formulas for other arithmetical functions.

1.101. If f is a completely multiplicative function then for all primes p,

$$f_p(X) = \frac{1}{1 - f(p)X} .$$

If g is a specially multiplicative function and if $B = B_g$, then for all primes p,

$$g_p(X) = \frac{1}{1 - g(p)X + B(p)X^2}$$

1.102. Let g be a multiplicative function and h a completely multiplicative function. If for all primes p,

$$g_p(X) = \frac{1}{1 - g(p)X + h(p)X^2} ,$$

then g is specially multiplicative and $h = B_g$.

Notes on Chapter 1

The early history of the theory of arithmetical functions is contained in the first volume of L. E. Dickson's monumental History of the Theory of Numbers [D]. Three chapters are relevant: Chapter V. "Euler's ϕ-function, generalizations; Farey series," Chapter X. "Sum and number of divisors" and Chapter XIX. "Inversion of functions; Möbius' function $\mu(n)$; numerical integrals and derivatives." The earliest document to which reference is made in the three chapters is a letter from L. Euler to D. Bernoulli written in 1741, so that our subject is approximately 245 years old.

The convolution operation played a prominent role from the very beginning. Many results from the early times involved the convolution of two or more particular arithmetical functions. Nowhere is this more evident than in the long list of arithmetical identities discovered by J. Liouville and published by him without proof in 1857 (see L. E. Dickson [D], pp. 285-286). Most of the identities in Exercises 1.48-1.59 are from that list.

Early in this century the Dirichlet convolution, as well as addition and multiplication, began to be viewed as binary operations on the set of arithmetical functions. In the work of M. Cipolla and E. T. Bell it was recognized that the arithmetical functions form a commutative ring with unity with respect to addition and convolution. An exposition of M. Cipolla's work was made by F. Pellegrino [53], and E. T. Bell's early results are contained in his paper [15], and in a summary of his work in his letter [28] to R. Vaidyanathaswamy, who was working along the same lines.

The study of the structure of the ring of arithmetical functions has continued, and we point out papers of L. Carlitz [52], [64], E. Cashwell and C. Everett [59] and H. N. Shapiro [72]. There are operations other than Dirichlet convolution which, together with addition, make the set of arithmetical functions into a ring. Some of them are discussed in Chapter 4. The Cauchy product, defined in the next chapter, is the multiplication in the rings considered by E. Cohen [52], [54]. A survey of various binary operations on the set of arithmetical functions was given by M. V. Subbarao [72].

The Möbius function first appeared in 1832 in the work of A. F. Möbius on the inversion of series of functions. J. H. Loxton and J. W. Sanders [80] discussed the subsequent development of applications of the Möbius function to the inversion of series and to other problems not of a number-theoretic nature. We shall return to this in the notes on Chapter 5.

Theorem 1.3 is simply the first in a long line of inversion theorems for arithmetical functions. The results in Exercises 1.15 and 1.25 are examples of others: the former was proved by P. J. McCarthy [68], and the latter is such a natural generalization of Theorem 1.3 that is is rediscovered as needed with some regularity. Inversion theorems were discussed by D. E. Daykin [64], [72], U. V. Satyanarayana [63], [65], T. M. Apostol [65], R. T. Hansen [80] and R. T. Hansen and L. G. Swanson [80].

It is a remarkable fact that both Theorem 1.3 and the Inclusion-Exclusion Principle are consequences of a very general theorem in the theory of lattices. It is that part of the theory of lattices which is the subject matter of Chapter 7. The Inclusion-Exclusion Principle is contained in Exercise 7.44. Some applications of that principle to the evaluation of various arithmetical functions were made by S. Pankajan [36].

The early history of the Jordan function J_k was summarized by L. E. Dickson [D], pp. 147-155, and it was pointed out there that R. D. von Sterneck proved that $J_k = H_k$: his proof is contained in Exercises 1.7, 1.42 and 1.43. The function ϕ_k in Exercise 1.37 was introduced by E. Cohen [49], and the results stated in Exercises 1.36-1.41 were proved in that paper and in other papers by the same author [56b], [58a], [59a].

Completely multiplicative functions were called linear functions by R. Vaidyanathaswamy, because of the form of their Bell series (see Exercise 1.101). Proposition 1.9 and the necessity of the condition in Proposition 1.8 were stated by R. Vaidyanathaswamy [27]. He observed also the necessity of the condition in Proposition 1.10, and the full result was stated by J. Lambek [66]. T. M. Apostol [71] wrote an exposition of the theory of completely multiplicative functions which included Propositions 1.8-1.10 and the results in Exercises 1.45 and 1.46. For some remarks concerning Exercise 1.45, see the paper of E. Langford [73]. The special case in Exercise 1.45 is due to R. Sivaramakrishnan [70], and the result in Exercise 1.44 to L. Carlitz [71].

The study of specially multiplicative functions, along the lines of Theorem 1.12, was carried out by R. Vaidyanathaswamy [30], [31]. He called them quadratic functions (see, again, Exercise 1.101), and the term "specially multiplicative function" was coined by D. H. Lehmer [59]. In his 1930 paper R. Vaidyanathaswamy derived the identity in Exercise 1.67: he called it the identical equation of the function f. He observed that for a specially multiplicative function the identical equation is the identity of Theorem 1.12(2). The identical equation is the single-variable case of a general result obtained by R. Vaidyanathaswamy in his 1931 paper for arithmetical functions of several variables. A proof of the identical equation in the manner suggested in Exercise 1.68 was given by A. A. Gioia [62].

The statement in Exercise 1.62 was also proved by R. Vaidyanathaswamy in his 1930 paper, as was the necessity of the condition in Exercise 1.61. The complete result of the latter exercise was obtained by T. B. Carroll and A. A. Gioia [75], and the case $k = 2$ independently by R. Sivaramakrishnan [76]. The identity in Exercise 1.63 is due to R. Vaidyanathaswamy [31]: another proof was given by J. Lambek [66].

In the introduction to his 1931 paper, R. Vaidyanathswamy stated that the paper arose from an effort to understand the identity

$$\sigma_k(mn) = \sum_{d|(m,n)} \sigma_k(m/d)\sigma_k(n/d)\mu(d)d^k \quad \text{for all } m, n,$$

"and to answer the converse problem suggested by it, of finding the most general function admitting an identity of this form" (p. 582). He was completely successful in his efforts, and proved Theorem 1.12. In fact, his Theorem XXXV is a much more general result, for arithmetical functions of several variables. The identity above in the case in which $k = 0$ was stated by S. Ramanujan [16a], and the complete identity was verified by S. Chowla [29]. The proof of Theorem 1.12 given in the text is from a paper by P. J. McCarthy [60e].

The similar problem for the restricted Busche-Ramanujan identity was also solved by R. Vaidyanathaswamy [31], in another result which holds for arithmetical functions of several variables (see his Theorem XXXVIII). His result, for functions of a single variable, form the content of Exercises 1.80-1.85. He introduced the kth convolute of an arithmetical function, and this notion was investigated in greater detail by H. Scheid [69] and by H. Scheid and R. Sivaramakrishnan [70]. The results in Exercises 1.79 and 1.86-1.88 were obtained by P. J. McCarthy [60d], [60e], [62].

Specially multiplicative functions arise naturally in the theroy of modular forms, which are the subject of a book by T. M. Apostol [A']. The values of certain specially multiplicative functions occur as the Fourier coefficients of modular forms, and in Chapter 6 of [A'] a result of E. Hecke is proved in which those modular forms giving rise in this way to specially multiplicative functions are characterized.

One function which arises in this way is Ramanujan's tau-function τ (not to be confused with the number-of-divisors function: it is only in these notes to Chapter 1 that τ denotes Ramanujan's tau-function). The function was introduced by S. Ramanujan [16b] by means of the identity

$$\sum_{n=1}^{\infty} \tau(n)x^n = x \prod_{k=1}^{\infty} (1 - x^k)^{24} .$$

The series and the product converge for $|x| < 1$. The definition of τ via a modular form is on page 20 of [A'], and on pages 92-93 there is an outline of a proof that τ is specially multiplicative, with $B_\tau = \zeta_{11}$. Thus,

$$\tau(p^{\alpha+1}) = \tau(p)\tau(p^\alpha) - p^{11}\tau(p^{\alpha-1})$$

for all primes p and all $\alpha \geq 1$, and for all m and n,

$$\tau(mn) = \sum_{d|(m,n)} \tau(m/d)\tau(n/d)\mu(d)d^{11},$$

which was noted by K. G. Ramanathan [43b], and

$$\tau(m)\tau(n) = \sum_{d|(m,n)} d^{11}\tau(mn/d^2).$$

Therefore, by the proof (that (4) $\Rightarrow$ (1)) of Theorem 1.12, $\tau = g_1 * g_2$ where each g_i is completely multiplicative and for each prime p,

$$g_i(p) = \frac{1}{2} (\tau(p) + (-1)^i \sqrt{\tau(p)^2 - 4p^{11}}).$$

It was conjectured by S. Ramanujan [16b] that the quantity under the radical is negative for each prime p: the conjecture was proved by P. Deligne (see [A'], p. 136).

G. H. Hardy [H] devoted a chapter to the tau-function, and an exposition of results concerning τ was written by F. van der Blij [50]. The latter has an extensive, albeit dated, list of references. In a recent paper, J. A. Ewell [84] obtained a formula for $\tau(n)$ in terms of the number of ways of writing an integer as a sum of sixteen squares.

The norm $N(f)$ of a multiplicative function f was defined by P. Kesava Menon [63] in a paper dealing with S. Ramanujan's tau-function. The norms of specially multiplicative functions were studied by R. Sivaramakrishnan [76].

The evaluation of the sum in Exercise 1.7 is generally attributed to P. Kesava Menon [65], but it is contained in a general result due to E. Cohen [59b]. This point will be made again in the notes on Chapter 2. The result has been proved and/or generalized many times, for example in papers by R. Nageswara Rao [66c], [72], R. Sivaramakrishnan [69], [74], S. Venkatramaiah [73], T. Venkataraman [74], V. Sita Ramaiah [79] and I. M. Richards [84].

The function ϕ_f of Exercise 1.21 was defined by P. Kesava Menon [67], and independently by H. Stevens [71]. This function, and others defined with respect to a polynomial or a set of polynomials, were studied by J. Chidambaraswamy [74], [76a], [76b], [79a].

The basic properties of the function Φ_k defined in Exercise 1.29 were set down by V. Klee [48], and for this reason it is generally referred to as Klee's function, even though it had been considered as far back as 1900 by F. Rogel (see L. E. Dickson [D], p. 134). A few years before V. Klee's note appeared, the function Φ_2 had been studied by E. K. Haviland [44]. Other results concerning Φ_k were obtained by P. J. McCarthy [58], K. Nageswara Rao [61b], U. V. Satyanarayana and K. Pattabhiramasastry [65] and A. C. Vasu [72b].

The functions τ_k and $\tau_{k,h}$ in Exercises 1.32 and 1.33 are defined by M. G. Beumer [62] and R. Sivaramakrishnan [68], respectively. The latter obtained the result relating $\tau_{k,h}$ and Klee's function. In fact, the function τ_k has been discovered a number of times (see L. E. Dickson [D], p. 135, p. 287 and p. 308).

The core function in Exercise 1.14 was introduced by S. Wigert [32], and it has been studied by others, including E. Cohen [60c] and D. Suryanarayana [72a]. The function δ_k of Exercise 1.20 was also studied by D. Suryanarayana [69b]. For the early work on Schemmel's function in Exercise 1.23 see the page devoted to it by L. E. Dickson, [D], p. 147. A generalization was made by K. Nageswara Rao [66b]. The generalizations of Dedekind's function given in Exercises 1.34 and 1.35 were defined by D. Suryanarayana [69a] and J. Hanumanthachari [72]. The (k,q)-integers and the associated arithmetical functions in Exercises 1.92-1.96, were introduced by M. V. Subbarao and V. C. Harris [66].

Finally, the Bell series in Exercises 1.97-1.102 were introduced by E. T. Bell [15]. Bell series for arithmetical functions of several variables were the principal tool used by R. Vaidyanathaswamy [31].

Chapter 2
Ramanujan Sums

If a and b are integers let

$$e(a,b) = e^{\frac{2\pi ia}{b}} .$$

Let n be an integer, positive, negative or zero, and let r be a positive integer. Consider the sum

$$c(n,r) = \sum_{(x,r)=1} e(nx,r) .$$

Usually, the sum is taken over all x such that $1 \le x \le r$ and $(x,r) = 1$, but it could be over any reduced residue system (mod r). This is because, if $x \equiv x'$ (mod r) then $e(nx,r) = e(nx',r)$. The sum $c(n,r)$ is called a Ramanujan sum. For fixed r, and with n restricted to the positive integers, we obtain an arithmetical function $c(\cdot,r)$. Some authors devote this function by c_r, so that $c_r(n) = c(n,r)$.

On the other hand, for fixed n we obtain an arithmetical function $c(n,\cdot)$. When $n = 0$ this is Euler's function,

$$c(0,r) = \phi(r) \quad \text{for all } r ,$$

and by the proposition that follows,

$$c(1,r) = \mu(r) \quad \text{for all } r .$$

The value of the sum is easy to determine.

Proposition 2.1. For all n and r ,

$$c(n,r) = \sum_{d|(n,r)} d\mu(r/d) .$$

Proof. By Exercise 2.1,

$$g(n,r) = \sum_{h=1}^{r} e(nh,r) = \begin{cases} r & \text{if } r|n \\ 0 & \text{if } r\nmid n . \end{cases}$$

By Lemma 1.4,

$$g(n,r) = \sum_{d|r} \sum_{(x,d)=1} e(nxr/d,r)$$

$$= \sum_{d|r} \sum_{(x,d)=1} e(nx,d)$$

$$= \sum_{d|r} c(n,d) .$$

Therefore, by the Möbius inversion formula,

$$c(n,r) = \sum_{d|r} g(n,d)\mu(r/d)$$

$$= \sum_{d|(n,r)} d\mu(r/d). \square$$

Corollary 2.2. For all n,r and s with $(r,s) = 1$,

$$c(n,rs) = c(n,r)c(n,s) .$$

If p is a prime and $\alpha \geq 1$, then by the formula of Proposition 2.1,

$$c(n,p^\alpha) = \begin{cases} p^\alpha - p^{\alpha-1} = p^\alpha(1-\frac{1}{p}) & \text{if } p^\alpha | n \\ -p^{\alpha-1} & \text{if } p^\alpha \nmid n \text{ but } p^{\alpha-1} | n \\ 0 & \text{otherwise .} \end{cases}$$

Thus, by Corollary 2.2, $c(n,r)$ is integer-valued.

Let g be a multiplicative function and h a completely multiplicative function, and consider the sum

$$f(n,r) = \sum_{d|(n,r)} h(d)g(r/d)\mu(r/d) ,$$

where n is an integer and r is a positive integer. The associated arithmetical function F is defined by

$$F(r) = f(0,r) \quad \text{for all } r .$$

For example, if $h = \zeta_1$ and $g = \zeta$ then

$$f(n,r) = c(n,r) \text{ and } F = \phi.$$

The function F is multiplicative, and if p is a prime and $\beta \geq 1$, then

$$F(p^\beta) = \sum_{j=0}^{\beta} h(p^j)g(p^{\beta-j})\mu(p^{\beta-j})$$

$$= h(p)^{\beta-1}(h(p) - g(p)).$$

Thus, $F(r) \neq 0$ for all r if and only if

(*) $h(p) \neq 0$ and $h(p) \neq g(p)$ for all primes p.

<u>Theorem 2.3</u>. If (*) holds then for all n and r,

$$f(n,r) = \frac{F(r)g(m)\mu(m)}{F(m)} \quad \text{where} \quad m = \frac{r}{(n,r)}.$$

Proof. If

$$(n_1,n_2) = (r_1,r_2) = (n_1,r_2) = (n_2,r_1) = 1$$

then (n_1,r_1) and (n_2,r_2) are relatively prime and

$$(n_1n_2,r_1r_2) = (n_1,r_1)(n_2,r_2).$$

Thus,

$$f(n_1 n_2, r_1 r_2) = f(n_1, r_1) f(n_2, r_2).$$

Therefore, if $n = p_1^{\alpha_1} \ldots p_t^{\alpha_t}$ and $r = p_1^{\beta_1} \ldots p_t^{\beta_t}$, where some of the α's and β's may equal zero, then

$$f(n,r) = \prod_{i=1}^{t} f(p_i^{\alpha_i}, p_i^{\beta_i}) .$$

The right-hand side of the formula of the theorem factors in the same way. Hence, it is sufficient to show that if p is a prime and $\alpha, \beta \geq 0$, then

$$f(p^\alpha, p^\beta) = \frac{F(p^\beta) g(p^\gamma) \mu(p^\gamma)}{F(p^\gamma)} \quad \text{where} \quad p^\gamma = \frac{p^\beta}{(p^\alpha, p^\beta)} .$$

We shall denote the right-hand side of this equation by $f'(p^\alpha, p^\beta)$.

Case 1: $\beta \leq \alpha$. Then $p^\gamma = 1$ and so

$$f'(p^\alpha, p^\beta) = F(p^\beta) = f(p^\alpha, p^\beta) .$$

Case 2: $\beta - 1 = \alpha$. Then $p^\gamma = p$ and so

$$f'(p^\alpha, p^\beta) =- \frac{F(p^\beta) g(p)}{F(p)}$$

$$= \frac{h(p)^{\beta-1}(h(p)-g(p))g(p)\mu(p)}{h(p)-g(p)}$$

$$= h(p^{\beta-1})g(p)\mu(p) = f(p^{\alpha},p^{\beta}) .$$

Case 3: $\beta - 1 > \alpha$. Then $\gamma \geq 2$ and so $f'(p^{\alpha},p^{\beta}) = 0$. Also, $f(p^{\alpha},p^{\beta}) = 0$. □

Corollary 2.4. For all n and r ,

$$c(n,r) = \frac{\phi(r)\mu(m)}{\phi(m)} \quad \text{where} \quad m = \frac{r}{(n,r)} .$$

Theorem 2.5. If (*) holds then for all n and r ,

$$F(r) \sum_{\substack{d|r \\ (n,d)=1}} \frac{h(d)}{F(d)} \mu(r/d) = \mu(r)f(n,r) .$$

It is sufficient to prove that the equality holds whenever n and r are powers of the same prime. The details are left as Exercise 2.16.

A special case of the identity in Theorem 2.5 is

$$\phi(r) \sum_{\substack{d|r \\ (n,d)=1}} \frac{d}{\phi(d)} \mu(r/d) = \mu(r)c(n,r)$$

$$= \mu(r) \sum_{d|(n,r)} d\mu(r/d) .$$

This is called the Brauer-Rademacher identity.

Let r be a positive integer. An arithmetical function f is said to be <u>periodic (mod r)</u> if $f(n) = f(n')$ whenever $n \equiv n' \pmod r$. Note that if f is periodic (mod d), where $d|r$, then f is periodic (mod r).

It is clear from Proposition 2.1 that $c(n,r) = c((n,r),r)$ for all n and r. Since $n \equiv n' \pmod r$ implies that $(n,r) = (n',r)$, it follows that the arithmetical function $c(\cdot,r)$ is periodic (mod r).

If the arithmetical functions f and g are periodic (mod r), their <u>Cauchy product</u> is the arithmetical function h defined by

$$h(n) = \sum_{n\equiv a+b \pmod r} f(a)g(b) .$$

The sum is over all solutions $\langle a, b \rangle$ (mod r) of the congruence $n \equiv X + Y \pmod r$. Note that the function h is also periodic (mod r).

Many of the useful properties of the Ramanujan sums are consequences of the fact that they are orthogonal with respect to the Cauchy product.

<u>Theorem 2.6</u>. For all n and r, if $d,e|r$ then

$$\sum_{n\equiv a+b \pmod r} c(a,d)c(b,e) = \begin{cases} rc(n,d) & \text{if } d = e \\ 0 & \text{if } d \neq e . \end{cases}$$

The proof depends on the following property of the exponential function.

<u>Lemma 2.7</u>. Let $d,e|r$, $1 \leq x \leq d$, $1 \leq y \leq e$ and $(x,d) = (y,e) = 1$. Then

$$\sum_{n\equiv a+b \pmod r} e(ax,d)e(by,e)$$

$$= \begin{cases} re(nx,d) & \text{if } d = e \text{ and } x = y \\ 0 & \text{otherwise.} \end{cases}$$

Proof. Let $r/d = d_1$ and $r/e = e_1$. The sum is equal to

$$\sum_{n\equiv a+b \pmod r} e(axd_1,r)e(bye_1,r)$$

$$= \begin{cases} re(nxd_1,r) = re(nx,d) & \text{if } xd_1 = ye_1 \\ 0 & \text{otherwise,} \end{cases}$$

by Exercise 2.18. However, $xd_1 = ye_1$ if and only if $xe = yd$, and since $(x,d) = (y,e) = 1$, this is true if and only if $x = y$ and $d = e$. □

Proof of Theorem 2.6. The sum is equal to

$$\sum_{\substack{(x,d)=1 \\ 1\le x\le d}} \sum_{\substack{(y,e)=1 \\ 1\le y\le e}} \sum_{n\equiv a+b \pmod r} e(ax,d)e(by,e) .$$

If $d \neq e$ the inner sum is equal to zero for all x and y, and if $d = e$ the inner summ is equal to zero unless $x = y$. Hence, if $d = e$ the triple sum is equal to

$$\sum_{\substack{(x,d)=1 \\ 1\le x\le d}} re(nx,d) = rc(n,d). \ \square$$

The orthogonality property can be put into a second, useful form.

Theorem 2.8. For all r, if $e_1, e_2 | r$ then

$$\sum_{d|r} c(r/d, e_1)c(r/e_2, d) = \begin{cases} r & \text{if } e_1 = e_2 \\ 0 & \text{if } e_1 \neq e_2 . \end{cases}$$

Proof. Let

$$S = \sum_{a+b \equiv 0 \pmod{r}} c(a, e_1)c(b, e_2)$$

$$= \sum_{a=1}^{r} c(a, e_1)c(-a, e_2)$$

$$= \sum_{d|r} \sum_{\substack{1 \le x \le d \\ (x,d)=1}} c(xr/d, e_1)c(-xr/d, e_2) ,$$

the last step by Lemma 1.4. For each x,

$$c(xr/d, e_1) = \sum_{(y,e_1)=1} e(yxr/d, e_1),$$

and $e(yxr/d, e_1) = e(yxr/e_1, d)$: hence $c(xr/d, e_1) = c(x'r/d, e_1)$ whenever $x \equiv x' \pmod{d}$. The same is true for $c(-xr/d, e_2)$. Thus,

$$S = \sum_{d|r} \sum_{x} c(xr/d, e_1)c(-xr/d, e_2),$$

where x runs over any reduced residue system (mod d). Furthermore, by

Exercise 1.39, we can assume that this reduced residue system (mod d) is contained in a reduced residue system (mod r), i.e., that $(x,r) = 1$ for each x. Then $c(\pm xr/d, e_i) = c(r/d, e_i)$ for $i = 1,2$ since $c(n,r) = c((n,r),r)$ for all n and r. Thus,

$$S = \sum_{d|r} c(r/d,e_1)c(r/d,e_2)\phi(d)$$

$$= \phi(e_2) \sum_{d|r} c(r/d,e_1)c(r/e_2,d) ,$$

by Exercise 2.20. On the other hand, by Theorem 2.6 with $n = 0$,

$$S = \begin{cases} r\phi(e_2) & \text{if } e_1 = e_2 \\ 0 & \text{if } e_1 \neq e_2 . \ \Box \end{cases}$$

If we let $e_1 = 1$ in Theorem 2.8 and write e for r/e_2, we see that if $e|r$ then

$$\sum_{d|r} c(e,d) = \begin{cases} r & \text{if } e = r \\ 0 & \text{if } e \neq r . \end{cases}$$

Other special cases of the equality in Theorem 2.8 are given in Exercises 2.22 - 2.24.

An arithmetical function f is called an <u>even function (mod r)</u> if

$$f((n,r)) = f(n) \quad \text{for all } n .$$

Note that if f has this property then it is periodic (mod r).

We have observed already that the function $c(\cdot,r)$ is an even function (mod r). In fact, the same is true of the $c(\cdot,d)$ for every divisor d of r. For,

$$\begin{aligned} c((n,r),d) &= c((n,r,d),d) \\ &= c((n,d),d) \\ &= c(n,d) \quad \text{for all } n . \end{aligned}$$

Theorem 2.9. If f is an even function (mod r) then f can be written uniquely in the form

$$f(n) = \sum_{d|r} \alpha(d)c(n,d) \qquad \text{for all } n .$$

The coefficients $\alpha(d)$ are given by

$$\begin{aligned} \alpha(d) &= \frac{1}{r}\sum_{e|r} f(r/e)c(r/d,e) \\ &= \frac{1}{r\phi(d)}\sum_{m=1}^{r} f(m)c(m,d) . \end{aligned}$$

Proof. First we shall verify the uniqueness of the representation by showing that if

$$\sum_{d|r} \alpha(d)c(n,d) = 0 \qquad \text{for all } n$$

then $\alpha(e) = 0$ for every divisor e of r. We have

$$0 = \sum_{a+b\equiv 0 \pmod r} \sum_{d|r} \alpha(d)c(a,d)c(b,e)$$

$$= \sum_{d|r} \alpha(d) \sum_{a+b\equiv 0 \pmod r} c(a,d)c(b,e)$$

$$= \alpha(e)r\phi(e)$$

by Theorem 2.6. Therefore, $\alpha(e) = 0$.

Now suppose that $\alpha(d)$ is given by the first of the two formulas. Then

$$\sum_{d|r} \alpha(d)c(n,d) = \frac{1}{r}\sum_{d|r} \sum_{e|r} f(r/e)c(r/d,e)c(n,d)$$

$$= \frac{1}{r}\sum_{e|r} f(r/e) \sum_{d|r} c(r/d,e)c(r/b,d) .$$

where $b = r/(n,r)$. By Theorem 2.8 the inner sum is equal to zero unless $e = b$, in which case it is equal to r. Thus,

$$\sum_{d|r} \alpha(d)c(n,d) = f(r/b) = f((n,r)) = f(n) .$$

Finally, let

$$S = \sum_{m=1}^{r} f(m)c(m,d) = \sum_{e|r} \sum_{\substack{1\le x\le e \\ (x,e)=1}} f(xr/e)c(xr/e,d) .$$

If $x \equiv x'$ (mod e) then $xr/e \equiv x'r/e$ (mod r): hence

$$f(xr/e)c(xr/e,d) = f(x'r/e)c(x'r/e,d) .$$

Thus,

$$S = \sum_{e|r} \sum_{x} f(xr/e)c(xr/e,d) ,$$

where for each e, x runs over <u>any</u> reduced residue system (mod e). Again by Exercise 1.39, we can assume that each x is relatively prime to r . Then, for each x , $(xr/e,r) = r/e$: hence $f(xr/e) = f(r/e)$. Likewise, $c(xr/e,d) = c(r/e,d)$ since $(xr/e,d) = (r/e,d)$. Thus,

$$\begin{aligned} S &= \sum_{e|r} f(r/e)c(r/e,d)\phi(e) \\ &= \phi(d) \sum_{e|r} f(r/e)c(r/d,e) \quad \text{by Exercise 2.20} \\ &= r\phi(d)\alpha(d) . \ \square \end{aligned}$$

The coefficients $\alpha(d)$ are called the <u>Fourier coefficients</u> of the function f .

Note that if α is any complex-valued function defined on the set of divisors of r , then the arithmetical function f defined by

$$f(n) = \sum_{d|r} \alpha(d)c(n,d) \quad \text{for all } n$$

is an even function (mod r) .

Theorem 2.10. An arithmetical function f is an even function (mod r) if and only if there is a complex-valued function g of two positive integer variables such that

$$f(n) = \sum_{d|(n,r)} g(d,r/d) \quad \text{for all } n .$$

In this case, for each divisor d of r,

$$\alpha(d) = \frac{1}{r}\sum_{e|\frac{r}{d}} g(r/e,e)e .$$

Proof. Certainly, a function f given as above for some g, is an even function (mod r). Conversely, suppose f is an even function (mod r). Then, for all n,

$$f(n) = \sum_{d|r} \alpha(d)c(n,d) = \sum_{d|r} \alpha(d) \sum_{e|(n,d)} e\mu(d/e)$$

$$= \sum_{e|(n,r)} e \sum_{\substack{d|r \\ e|d}} \alpha(d)\mu(d/e)$$

$$= \sum_{e|(n,r)} e \sum_{D|\frac{r}{e}} \alpha(De)\mu(D) .$$

Thus, f has the required form with

$$g(a,b) = a\sum_{D|b} \alpha(Da)\mu(D) .$$

Furthermore, since

$$\sum_{d|r} \frac{1}{r} \sum_{e|\frac{r}{d}} g(r/e,e)e\ c(n,d)$$

$$= \sum_{e|r} \frac{1}{r} g(r/e,e)e \sum_{d|\frac{r}{e}} c(n,d)$$

$$= \sum_{\substack{e|r \\ \frac{r}{e}|n}} g(r/e,e) = \sum_{d|(n,r)} g(d,r/d) = f(n) ,$$

the formula for $\alpha(d)$ follows. □

Because an even function (mod r) is periodic (mod r) we can consider the Cauchy product of two such functions.

Proposition 2.11. Let f and g be even functions (mod r), with Fourier coefficients $\alpha(d)$ and $\alpha'(d)$, respectively. Then their Cauchy product h is an even function (mod r) with Fourier coefficients $r\alpha(d)\alpha'(d)$.

Proof. For all n ,

$$h(n) = \sum_{n\equiv a+b \pmod r} \sum_{d|r} \alpha(d)c(a,d) \sum_{e|r} \alpha'(d)c(b,e)$$

$$= \sum_{d|r} \sum_{e|r} \alpha(d)\alpha'(e) \sum_{n\equiv a+b \pmod r} c(a,d)c(b,e)$$

$$= \sum_{d|r} r\alpha(d)\alpha'(d)c(n,d)$$

by Theorem 2.6. □

In the rest of this chapter we consider some special situations to which the preceding results can be applied.

For all n and r

$$\delta((n,r)) = \begin{cases} 1 & \text{if } (n,r) = 1 \\ 0 & \text{if } (n,r) \neq 1 . \end{cases}$$

For a fixed r, $\delta((\cdot,r))$ is an even function (mod r). By Theorem 2.9, its Fourier coefficients are

$$\begin{aligned} \alpha'(d) &= \frac{1}{r} \sum_{e|r} \delta((r/e,r)) c(r/d,e) \\ &= \frac{1}{r} \sum_{\substack{e|r \\ (\frac{r}{e},r)=1}} c(r/d,e) \\ &= \frac{1}{r} c(r/d,r) = \frac{1}{r} \frac{\phi(r)\mu(d)}{\phi(d)} , \end{aligned}$$

by Corollary 2.4.

Now let f be an even function (mod r) with Fourier coefficients $\alpha(d)$. Then, by Proposition 2.11, for all n,

$$\begin{aligned} \sum_{(b,r)=1} f(n-b) &= \sum_{n \equiv a+b \,(\mathrm{mod}\ r)} f(a)\delta((b,r)) \\ &= \phi(r) \sum_{d|r} \frac{\alpha(d)\mu(d)}{\phi(d)} c(n,d) . \end{aligned}$$

In particular, if $f = c(\cdot,r)$,

$$\sum_{(b,r)=1} c(n-b,r) = \mu(r)c(n,d)$$

(for a more general result, see Exercise 2.25). On the other hand,

$$\sum_{(b,r)=1} c(n-b,r) = \sum_{(b,r)=1} \sum_{d|(n-b,r)} d\mu(r/d)$$

$$= \sum_{(b,r)=1} \sum_{\substack{d|r \\ n\equiv b \pmod{d}}} d\mu(r/d)$$

$$= \sum_{d|r} d\mu(r/d) \sum_{\substack{(b,r)=1 \\ b\equiv n \pmod{d}}} 1 .$$

For each d, the inner sum is over all elements b of a reduced residue system (mod r) such that $b \equiv n \pmod{d}$. If $(n,d) > 1$ there are no such elements. If $(n,d) = 1$ then by Exercise 1.39 there are exactly $\phi(r)/\phi(d)$ such elements. Thus,

$$\sum_{(b,r)=1} c(n-b,r) = \phi(r) \sum_{\substack{d|r \\ (n,d)=1}} \frac{d\mu(r/d)}{\phi(d)} .$$

Therefore, for all n,

$$\mu(r)c(n,r) = \phi(r) \sum_{\substack{d|r \\ (n,d)=1}} \frac{d\mu(r/d)}{\phi(d)} :$$

this is the Brauer-Rademacher identity, which was stated earlier as a special case of Theorem 2.5.

More generally, for an arbitrary f,

$$\sum_{(b,r)=1} f(n-b) = \phi(r) \sum_{d|r} \frac{\alpha(d)}{\phi(d)} \Big(\phi(d) \sum_{\substack{e|d \\ (n,e)=1}} \frac{e\mu(d/e)}{\phi(e)}\Big)$$

$$= \phi(r) \sum_{d|r} \alpha(d) \sum_{\substack{e|d \\ (n,e)=1}} \frac{e\mu(d/e)}{\phi(e)}$$

$$= \phi(r) \sum_{\substack{e|r \\ (n,e)=1}} \frac{e}{\phi(e)} \sum_{D|\frac{r}{e}} \alpha(De)\mu(D) \ .$$

Since an even function (mod r) can be defined by choosing the Fourier coefficients $\alpha(d)$ quite arbitrarily, we have the following result.

Proposition 2.12. Let r be a positive integer and let α be a complex-valued function on $\{1,\ldots,r\}$. Then, for all n,

$$\sum_{d|r} \frac{\alpha(d)\mu(d)}{\phi(d)} c(n,d) = \sum_{\substack{d|r \\ (n,d)=1}} \frac{d}{\phi(d)} \sum_{D|\frac{r}{e}} \alpha(Dd)\mu(D) \ .$$

Note that if

$$f(n) = \sum_{d|r} \alpha(d)c(n,d) \qquad \text{for all } n \ ,$$

then the right-hand side of the identity in Proposition 2.12 is

$$\sum_{\substack{d|r \\ (n,d)=1}} \frac{g(d,r/d)}{\phi(d)} ,$$

where g is the function related to f as in Theorem 2.10.

The identity in Proposition 2.12 can be thought of as a generalized Brauer-Rademacher identity. If α is chosen judiciously some identities result involving well-known arithmetical functions. Let h be an arbitrary arithmetical function and for each divisor d of r, let $\alpha(d) = h(r/d)$. Then

$$\sum_{D|\frac{r}{d}} \alpha(Dd)\mu(D) = \sum_{D|\frac{r}{d}} h((r/d)/D)\mu(D) = (h * \mu)(r/D) .$$

Thus, for all n,

$$\sum_{d|r} \frac{h(r/d)\mu(d)}{\phi(d)} c(n,d) = \sum_{\substack{d|r \\ (n,d)=1}} \frac{d}{\phi(d)} (h * \mu)(r/d) .$$

For example, if $h = \zeta_k$ then $h * \mu = J_k$: hence, for all n,

$$r^k \sum_{d|r} \frac{\mu(d)}{d^k\phi(d)} c(n,d) = \sum_{\substack{d|r \\ (n,d)=1}} \frac{dJ_k(r/d)}{\phi(d)} .$$

Or, if $h = \zeta$, so that $h * \mu = \delta$, then for all n,

$$\sum_{d|r} \frac{\mu(d)}{\phi(d)} c(n,d) = \begin{cases} \frac{r}{\phi(r)} & \text{if} \quad (n,r) = 1 \\ 0 & \text{if} \quad (n,r) > 1 . \end{cases}$$

In particular, with $n = 1$,

$$\sum_{d|r} \frac{|\mu(d)|}{\phi(d)} = \frac{r}{\phi(r)} .$$

Exercises for Chapter 2

2.1 For all r , and for every integer n , positive, negative or zero,

$$\sum_{h=1}^{r} e(nh,r) = \begin{cases} r & \text{if} \quad r|n \\ 0 & \text{if} \quad r\nmid n . \end{cases}$$

2.2. If $(mr,ns) = 1$ then

$$c(mn,rs) = c(m,r)c(n,s) .$$

2.3. If $(r,s) = 1$ then for all t ,

$$c(n,rt)c(n,st) = c(n,t)c(n,rst) .$$

2.4. If $(m,n) = 1$ then for all q ,

$$c(mq,r)c(nq,r) = c(q,r)c(mnq,r) .$$

In particular, if $(m,n) = 1$ then

$$c(m,r)c(n,r) = \mu(r)c(mn,r) .$$

2.5. If n and r are positive integers then

$$c(n,r)c(r,n) = \phi((n,r))c((n,r),[n,r]) .$$

2.6. If $(n,r) = 1$ then

$$c(mn,r) = c(m,r) ,$$

and

$$c(n,rs) = \mu(r)c(n,s) .$$

2.7. For all n and all even r,

$$\sum_{d|r} (-1)^d \, c(n,r/d) = \begin{cases} r & \text{if } n = r/2 \\ 0 & \text{otherwise} . \end{cases}$$

2.8. (See Exercises 1.1 and 1.2.) For all $n \neq 0$ and all r,

$$\sum_{j=1}^{r} c(n,j)\left[\frac{r}{j}\right] = \sum_{\substack{1 \le d \le r \\ d|n}} 1 .$$

Taking $n = r!$ this becomes

$$\sum_{j=1}^{r} \phi(j)\left[\frac{r}{j}\right] = \frac{1}{2}\, n(n+1) ,$$

and with $n = r! + 1$,

$$\sum_{j=1}^{r} \mu(j)\left[\frac{r}{j}\right] = 1 .$$

2.9. If $r > 1$ then

$$\sum_{d|r} c(d,r) = \phi(r,1) ,$$

where $\phi(\cdot,1)$ is Schemmel's function, defined in Exercise 1.23.

2.10. For all r ,

$$\sum_{d|r} c(d,r/d) = \begin{cases} r^{\frac{1}{2}} & \text{if } r \text{ is a square} \\ 0 & \text{otherwise} . \end{cases}$$

2.11. For all n and r ,

$$\sum_{d|n} c(d,r) = \sum_{d|(n,r)} \mu(r/d)\tau(n/d)d .$$

2.12. For all n (positive) and r ,

$$\sum_{d|r} \sum_{e|n} c(e,d) = \begin{cases} \tau(n/r)r & \text{if } r|n \\ 0 & \text{otherwise} . \end{cases}$$

2.13. For all n and r ,

$$\sum_{d|(n,r)} c(n/d,r/d) = \sum_{d|(n,r)} \sigma(d)\mu(r/d) .$$

2.14. For all n and r,

$$\sum_{(t,r)=1} c(nt,r) = \phi(r)c(n,r) .$$

2.15. For all n,

$$\sum_{\substack{d|n \\ (d,n/d)=1}} c(n/d,d) = \begin{cases} 1 & \text{if } n = 1 \text{ or if for all primes } p , \\ & p|n \text{ implies } p^2|n \\ 0 & \text{otherwise} . \end{cases}$$

2.16. Prove Theorem 2.5. (Hint: show first that if p is a prime and $\alpha \geq 1$, then $h(p^\alpha)/F(p^\alpha)$ is independent of α.)

2.17. If k is a positive integer then for all n and r,

$$J_k(r) \sum_{\substack{d|r \\ (n,d)=1}} \frac{d^k}{J_k(d)} \mu(r/d) = \mu(r) \sum_{d|(n,r)} d^k \mu(r/d) .$$

2.18. If $1 \leq s, t \leq r$ then for all n,

$$\sum_{n \equiv a+b \pmod r} e(as,r)e(bt,r) = \begin{cases} re(ns,r) & \text{if } s = t \\ 0 & \text{if } s \neq t \end{cases}$$

2.19. If the arithmetical function f is periodic (mod r) then there are complex numbers $c_0, \ldots, c_{r-1}$ uniquely determined by f, such that

$$f(n) = \sum_{j=0}^{r-1} c_j e(nj,r) \quad \text{for all } n .$$

In fact,

$$c_j = \frac{1}{r} \sum_{k=1}^{r} f(k)e(-kj,r) .$$

If g is also periodic (mod r) and

$$g(n) = \sum_{j=0}^{r-1} c_j' e(nj,r) \quad \text{for all } n ,$$

and if h is the Cauchy product of f and g, then

$$h(n) = r \sum_{j=0}^{r-1} c_j c_j' e(nj,r) \quad \text{for all } n .$$

2.20. In the notation of the paragraph preceding Theorem 2.3, if d and e are divisors of r then for all n,

$$F(d)f(nr/d,e) = F(e)f(nr/e,d) .$$

In particular,

$$\phi(d)c(nr/d,e) = \phi(e)c(nr/e,d) .$$

2.21. Theorem 2.8 can be proved directly, without going through Theorem 2.6. (Hint: use Exercise 2.2, and then prove the equality when r is a power of a prime. Begin by using the equation in the Exercise 2.20.)

2.22. For all r and s,

$$\sum_{n=1}^{qr} c(n,r)c(n,s) = \begin{cases} r\phi(r) & \text{if } r = s \\ 0 & \text{if } r \neq s . \end{cases}$$

2.23. For all n and r,

$$\sum_{d|r} c(r/d,r)c(n,d) = \begin{cases} r & \text{if } (n,r) = 1 \\ 0 & \text{if } (n,r) > 1 . \end{cases}$$

2.24. For every r, if $e|r$ then

$$\sum_{d|r} c(r/d,e)\phi(d) = \begin{cases} r & \text{if } e = 1 \\ 0 & \text{if } e \neq 1 . \end{cases}$$

and

$$\sum_{d|r} c(r/d,e)\mu(d) = \begin{cases} r & \text{if } e = r \\ 0 & \text{if } e \neq r . \end{cases}$$

2.25. Let h be an arithmetical function and define the function g of two positive integer variables by

$$g(n,r) = \sum_{d|(n,r)} h(d) \quad \text{for all } n \text{ and } r .$$

For each r, $g(\cdot,r)$ is an even function (mod r), and

$$\sum_{n\equiv a+b \pmod r} g(a,r)c(b,r) = h(r)c(n,r) \quad \text{for all } n .$$

2.26. Let g be a function of two positive integer variables. There exists an arithmetical function G such that $g(n,r) = G((n,r))$ for all n and r if and only if there is an arithmetical function h such that

$$g(n,r) = \sum_{d|(n,r)} h(d) \quad \text{for all } n \text{ and } r .$$

2.27. This exercise and the two that follow are special cases of Exercise 2.25. If d is a divisor of r then for all n,

$$\sum_{\substack{a=1 \\ (a,r)=d}}^{r} c(n-a,r) = \mu(r/d)c(n,r) .$$

2.28. For all n,

$$\sum_{\substack{a=1 \\ (a,r) \text{ a square}}}^{r} c(n-a,r) = \lambda(r)c(n,r) .$$

2.29. For all n ,

$$\sum_{\substack{a=1 \\ (a,r) \text{ squarefree}}}^{r} c(n-a,r) = \begin{cases} \mu(r^{\frac{1}{2}})c(n,r) & \text{if } r \text{ is a square} \\ 0 & \text{otherwise .} \end{cases}$$

2.30. Let s be an integer, positive, negative or zero (or even an arbitrary real number). Let h be a complex-valued function of two positive integer variables, and for all n and r let

$$f_s(n,r) = \sum_{d|(n,r)} \frac{h(d,r/d)}{d^s} , \quad f'_s(n,r) = \sum_{d|(n,r)} \frac{h(r/d,d)}{(r/d)^s}$$

If r is fixed then for all n ,

$$f_s(n,r) = \sum_{d|r} f'_{s+1}(r/d,r)c(n,d) .$$

2.31. Let s be as in Exercise 2.30 and define the arithmetical function β_s by

$$\beta_s(r) = \sum_{d|r} d^s\lambda(r/d) \quad \text{for all } r .$$

Then, for all n and r ,

$$\sum_{d|(n,r)} d^s\lambda(r/d) = \sum_{d|r} d^{s-1}\beta_{s-1}(r/d)c(n,d)$$

2.32. If r is a positive integer then for all n ,

$$\sum_{d|(n,r)} d\lambda(r/d) = \sum_{\substack{d|r \\ \frac{r}{d} \text{ a square}}} c(n,d) .$$

In particular, for all r ,

$$\lambda(r) = \sum_{d^2|r} \mu(r/d^2) \; , \; \beta(r) = \sum_{d^2|r} \phi(r/d^2) .$$

these identities were obtained in Chapter 1.

2.33. Let g be an arithmetical function and let the function h of the preceding exercise be given by $h(a,b) = g(b)$ for all a and b . If s is an integer, positive, negative or zero then for all n and r , let

$$G_s(n,r) = \sum_{d|(n,r)} d^s g(r/d) \; , \; G_s'(r) = G(0,r) .$$

If r is fixed then for all n ,

$$G_s(n,r) = \sum_{d|r} d^{s-1} G_{s-1}'(r/d) c(n,d) .$$

In particular, for all r ,

$$G_s'(r) = \sum_{d|r} d^{s-1} G_{s-1}'(r/d) \phi(d) .$$

(The latter is the identity $\zeta_{s-1}(\zeta_1 * \mu) * G_{s-1}' = \zeta_s * \zeta_{s-1}\mu * \zeta_{s-1} * g = \zeta_s * g = G_s'$, where we have used Propositions 1.8 and 1.10.) This identity contains many of those given in exercises in Chapter 1.

2.34. If k is a nonegative integer then for all n and r,

$$\frac{\sigma_k((n,r))}{(n,r)^k} = \frac{1}{r^{k+1}} \sum_{d|r} \sigma_{k+1}(r/d)c(n,d) .$$

(Hint: let $g = \zeta$ in Exercise 2.33.) In particular, for all r,

$$\sum_{d|r} \sigma_{k+1}(r/d)\phi(d) = r\sigma_k(r) .$$

2.35. If k is a positive integer then for all r,

$$\sum_{d|r} d^k J_k(r/d)\phi(d) = J_{k+1}(r) .$$

(Hint: let $g = \mu$ in Exercise 2.33.)

2.36. If s and t are positive integers then for all r,

$$\sum_{d|r} d^s \rho_{s,t}(r/d)\phi(d) = \rho_{s+1,t}(r)$$

(see Exercise 1.89 for the definition of Gegenbauer's function $\rho_{s,t}$). (Hint: take $g = \nu_t$ in Exercise 2.33.)

2.37. Let k be a nonnegative integer and r a positive integer, and define the arithmetical function $\tau_k(\cdot,r)$ by

$$\tau_k(n,r) = r^k \sum_{d|(n,r)} d\, J_{k+1}(r/d) \quad \text{for all } n ,$$

Then, for all n ,

$$\tau_k(n,r) = r^{2k+1} \sum_{d|r} \frac{1}{d^{k+1}} c(n,d) \quad \text{for all} \quad n .$$

(Hint: let $g = \zeta_k J_{k+1}$ in Exercise 2.33.)

2.38. Let h and k be nonnegative integers and let r and s be relatively prime positive integers. Then, for all n ,

$$\tau_k(n,rs) = \tau_k(n,r)\tau_k(n,s)$$

and

$$\sum_{n \equiv a+b \pmod r} \tau_h(a,r)\tau_k(b,r) = \tau_{h+k+1}(n,r) .$$

2.39. For all r ,

$$\frac{1}{r} \sum_{d|r} \frac{\tau_0(r/d,r)\,|\mu(d)|}{\phi(d)} = \tau(r) .$$

2.40. Let k and r be positive integers and define the arithmetical function f by

$$f(n) = (n,r)^k .$$

f is an even function (mod r) and its Fourier coefficients are $\alpha(d) = \frac{1}{r^k} \tau_{k-1}(r/d,r)$. Therefore,

$$\sum_{(b,r)=1} (n - b,r)^k = \frac{\phi(r)}{r^k} \sum_{d|r} \frac{\tau_{k-1}(r/d,r)\mu(d)}{\phi(d)} c(n,d) .$$

In particular,

$$\sum_{(b,r)=1} (b - 1,r) = \frac{\phi(r)}{r} \sum_{d|r} \frac{\tau_0(r/d,r)|\mu(d)|}{\phi(d)} .$$

Combining this with Exercise 2.39 yields the result of Exercise 1.7.

2.41. If r is a positive integer then for all n,

$$\sum_{d|r} \frac{d^k\sigma_k(r/d)\mu(d)}{\phi(d)} c(n,d) = \sum_{\substack{d|r \\ (n,d)=1}} \frac{d^{k+1}}{\phi(d)} .$$

In particular,

$$\sum_{d|r} \frac{d^k\sigma_k(r/d)|\mu(d)|}{\phi(d)} = \sum_{d|r} \frac{d^{k+1}}{\phi(d)}$$

and

$$\sum_{d|r} \frac{\sigma(r/d)|\mu(d)|}{\phi(d)} = r \sum_{d|r} \frac{1}{\phi(d)} .$$

(Hint: the final identity is obtained by taking $k = -1$ and writing out the sum.)

2.42. If r is a positive integer then for all n,

$$\sum_{d|r} \frac{d^k J_k(r/d)\mu(d)}{\phi(d)} c(n,d) = \sum_{\substack{d|r \\ (n,d)=1}} \frac{d^{k+1}\mu(r/d)}{\phi(d)} .$$

In particular,

$$\sum_{d|r} \frac{d^k J_k(r/d)|\mu(d)|}{\phi(d)} = \sum_{d|r} \frac{d^{k+1}\mu(r/d)}{\phi(d)} .$$

2.43. Let g be a multiplicative function and k a positive integer. For all n and r let

$$f_k(n,r) = \sum_{d|(n,r)} \frac{g(d)\mu(d)}{d^k} , \quad F_k(r) = f_k(0,r) .$$

If $g(p) \neq p^{k+1}$ for all primes p then

$$f_k(n,r) = F_{k+1}(r) \sum_{d|r} \frac{g(d)\mu(d)}{d^{k+1} F_{k+1}(d)} c(n,d) .$$

(Hint: use Theorem 2.3.)

2.44. If k is a positive integer then for all n and r,

$$\frac{J_k((n,r))}{(n,r)^k} = \frac{J_{k+1}(r)}{r^{k+1}} \sum_{d|r} \frac{\mu(d)}{J_{k+1}(d)} c(n,d) .$$

2.45. If k is a positive integer then for all n and r,

$$\sum_{d|(n,r)} \frac{|\mu(d)|d}{J_k(d)} = \frac{r^k}{J_k(r)} \sum_{d|r} \frac{|\mu(d)|}{d^k} c(n,d) .$$

2.46. If $k \geq 2$ is an integer then for all n and r,

$$\sum_{d|(n,r)} \frac{\mu(d)\phi(d)d}{J_k(d)} = \frac{r\, J_{k-1}(r)}{J_k(r)} \sum_{d|r} \frac{\mu(d)\phi(d)}{J_{k-1}(d)d} c(n,d) .$$

2.47. If k is a positive integer then for all n and r,

$$\sum_{d|(n,r)} \frac{|\mu(d)|\phi(d)}{J_k(d)} = \frac{J_{k+1}(r)}{rJ_k(r)} \sum_{d|r} \frac{|\mu(d)|\phi(d)}{J_{k+1}(d)} c(n,d) .$$

2.48. If f and g are even functions (mod r) with Fourier coefficients $\alpha(d)$ and $\beta(d)$, respectively, then for all r,

$$r \sum_{d|r} \alpha(d)\beta(d)\phi(d) = \sum_{d|r} f(r/d)g(r/d)\phi(d) .$$

2.49. If h and k are positive integers then for all r

$$r \sum_{d|r} d^{h+k} \sigma_h(r/d)\sigma_k(r/d)\phi(d) = \sum_{d|r} \sigma_{h+1}(r/d)\sigma_{k+1}(r/d)\phi(d) .$$

(Hint: use Exercise 2.48.)

2.50. Let f, B and g be as in Exercise 1.79. Let r be a positive integer and let

$$G(n,r) = \sum_{d|(N,r)} g(d)\mu(r/d) \quad \text{for all } N .$$

Then, for all m, n and N,

$$\sum_{d|(m,n)} B(d)f(m/d)f(n/d)G(N,d)$$

$$= \sum_{d|(m,n,N)} g(d)B(d)f(mn/d^2) .$$

In particular,

$$\sum_{d|(m,n)} B(d)f(m/d)f(n/d)c(N,d)$$

$$= \sum_{d|(m,n,N)} dB(d)f(mn/d^2) .$$

More particularly, for a nonnegative integer h,

$$\sum_{d|(m,n)} d\,\sigma_h(m/d)\sigma_h(n/d)c(N,d) = \sum_{d|(m,n,N)} d^{h+1}\sigma_h(mn/d^2) .$$

2.51. Let k be a positive integer. For all n (positive, negative or zero) and all r let

$$c_k(n,r) = \sum_{(x,r^k)_k=1} e(nx,r^k) .$$

the sum is over all x is an arbitrary reduced (r,k)-residue system. (See Exercise 1.36: it and those that follow it in Chapter 1 will be used in the sequence of exercises that begin with this one.) $c_k(n,r)$ is called a generalized Ramanujan sum. For all n and r ,

$$c_k(n,r) = \sum_{d^k|(n,r^k)} d^k\mu(r/d) .$$

In particular,

$$c_k(0,r) = J_k(r) \quad , \quad c_k(1,r) = \mu(r) .$$

2.52. For all n, r and s with $(r,s) = 1$,

$$c_k(n,rs) = c_k(n,r)c_k(n,s) ,$$

and for all primes p and all $\alpha \geq 1$,

$$c_k(n,p^\alpha) = \begin{cases} p^{\alpha k} - p^{(\alpha-1)k} & \text{if } p^{\alpha k}|n \\ -p^{(\alpha-1)k} & \text{if } p^{(\alpha-1)k}|n \text{ and } p^{\alpha k}\nmid n \\ 0 & \text{otherwise .} \end{cases}$$

2.53. For all n and r , if $d, e|r$, then

$$J_k(d)c_k(n(r/d)^k,e) = J_k(e)c_k(n(r/e)^k,d) .$$

2.54. For all n and r,

$$c_k(n,r) = \sum_{\substack{d \mid r^k \\ (d,r^k)_k = 1}} c(n,r^k/d) .$$

2.55. For all n and r,

$$c_k(n,r) = \frac{J_k(r)\mu(m)}{J_k(m)} , \quad m^k = \frac{r^k}{(n,r^k)_k} .$$

2.56. For all n and r, if $d, e \mid r$ then

$$\sum_{n \equiv a+b \pmod{r^k}} c_k(a,d)c_k(b,e) = \begin{cases} r^k c_k(n,d) & \text{if } d = e \\ 0 & \text{otherwise} . \end{cases}$$

(Hint: Formulate and prove the proper analogue of Exercise 2.18.)

2.57. For all r, if $e_1, e_2 \mid r$ then

$$\sum_{d \mid r} c_k((r/d)^k,e_1)c_k((r/e_2)^k,d) = \begin{cases} r^k & \text{if } e_1 = e_2 \\ 0 & \text{if } e_1 \neq e_2 . \end{cases}$$

2.58. Analogues of the identities in Exercises 2.22 - 2.24 hold for $c_k(n,r)$.

2.59. Let r be a positive integer. An arithmetical function f is an <u>(r,k)-even function</u> if $f((n,r^k)_k) = f(n)$ for all n. $c_k(\cdot,r)$ is such a function. If f is an (r,k)-even function then f can be written uniquely in the form

$$f(n) = \sum_{d|r} \alpha(d) c_k(n,d) \qquad \text{for all } n ,$$

where

$$\alpha(d) = \frac{1}{r^k} \sum_{e|r} f((r/e)^k) c_k((r/d)^k, e)$$

$$= \frac{1}{r^k J_k(d)} \sum_{m=1}^{r^k} f(m) c_k(m,d)$$

2.60. An arithmetical function f is an (r,k)-even function if and only if there is a function g of two positive integer variables such that for all n,

$$f(n) = \sum_{d^k | (n,r^k)} g(d, r/d) .$$

In this case,

$$\alpha(d) = \frac{1}{r^k} \sum_{e|\frac{r}{d}} g(r/e, e) e^k .$$

2.61. If, for all n,

$$f(n) = \sum_{d|r} \alpha(d) c_k(n,d) , \quad g(n) = \sum_{d|r} \beta(d) c_k(n,d) ,$$

then

$$\sum_{n \equiv a+b \pmod{r^k}} f(a)f(b) = r^k \sum_{d|r} \alpha(d)\beta(d)c_k(n,d) .$$

2.62. Let r and k be positive integers, and for all n let

$$c^{(k)}(n,r) = \sum_{(x_1,\ldots,x_k,r)=1} e(n(x_1 + \ldots + x_k), r) ,$$

where the sum is over all ordered k-tuples $\langle x_1, \ldots, x_k \rangle \pmod{r}$. For all n and r,

$$\sum_{d|r} c^{(k)}(n,d) = \begin{cases} r^k & \text{if } r|n \\ 0 & \text{if } r\nmid n , \end{cases}$$

$$c^{(k)}(n,r) = \sum_{d|(n,r)} d^k \mu(r/d) .$$

2.63. For all n and r,

$$c^{(k)}(n,r) = \frac{J_k(r)\mu(m)}{J_k(m)} , \quad m = \frac{r}{(n,r)} .$$

2.64. For integers n and r, with $r > 0$, let

$$B(n,r) = \sum e(nx,r),$$

where the sum is over all integers x such that $1 \le x \le r$ and (x,r) if a square. For all n and r,

$$\sum_{\substack{d|r \\ d \text{ squarefree}}} B(n,r/d) = \begin{cases} r & \text{if } r|n \\ 0 & \text{Otherwise,} \end{cases}$$

and

$$B(n,r) = \sum_{d|(n,r)} d\lambda(r/d) = \sum_{d^2|r} c(n,r/d^2) \quad \text{(see Exercise 2.32).}$$

2.65. Continuing Exercise 2.64,

$$B(1,r) = \lambda(r) \quad \text{and} \quad B(0,r) = \beta(r) \quad \text{for all } r.$$

For all n and r,

$$B(n,r) = \lambda(r/(n,r))\beta((n,r)).$$

β is the function defined near the end of Chapter 1.

2.66. If F is the function defined just before Theorem 2.3, and if (*) holds, then for all n and r,

$$\sum_{\substack{d|r \\ (n,d)=1}} \frac{g(d)|\mu(d)|}{F(d)} = \frac{h(r)F((n,r))}{F(r)h((n,r))} .$$

In particular, if k is a positive integer then for all n and r,

$$\sum_{\substack{d|r \\ (n,d)=1}} \frac{|\mu(d)|}{J_k(d)} = \frac{d^k J_k((n,r))}{J_k(d)(n,r)^k} .$$

When $k = n = 1$, this is the identity at the very end of the text of this chapter.

2.67. Let g and h be multiplicative functions and consider the sum

$$f(n,r) = \sum_{d|(n,r)} h(d)g(r/d)\mu(r/d),$$

where n and r are integers and r is positive. Let $F(r) = f(0,r)$ for all r, and assume that $F(r) \neq 0$ for all r. Assume also that $g(p) \neq 0$ for all primes p. If the formula of Theorem 2.3 holds for $f(n,r)$ then h is completely multiplicative.

2.68. Let $f(n,r)$ and F be as in Exercise 2.67. Assume that $h(p) \neq 0 \neq g(p)$ for all primes p. If the identity of Theorem 2.5 holds for $f(n,r)$ and F then h is completely multiplicative. (The assumption that $h(p) \neq 0$ for all primes p can be replaced by the assumption that for each prime p, if $h(p) = 0$ then $h(p^\alpha) = 0$ for all $\alpha \geq 1$.)

2.69. Let F be the arithmetical function defined in Exercise 2.67, and assume that $g(p) \neq 0$ for all primes p. If the identity of Exercise 2.66 holds for F then h is completely multiplicative.

2.70. If A is the $n \times n$ matrix $[c(i,j)]$, $1 \leq i,j \leq n$, then $\det A = n!$. (Hint: write

$$\sum_{d|r} c(m,d) = \begin{cases} r & \text{if } r|m \\ 0 & \text{if } r\nmid m \end{cases}$$

$1 \leq m,r \leq n$, in matrix form.)

2.71. Let f be a function of two positive integer variables such that for $r = 1,\ldots,n$, $f(\cdot,r)$ is an even function (mod r) with Fourier coefficients $\alpha(d,r)$. If A is the $n \times n$ matrix $[f(i,j)]$, $1 \leq i,j \leq n$, then $\det A = n!\, \alpha(1,1) \ldots \alpha(n,n)$.

2.72. Let f be as in Exercise 2.71, and let g be a function of two positive integer variables such that for $r = 1,\ldots,n$,

$$f(m,r) = \sum_{d|r} g(d,r/d) \quad \text{for all } m$$

(the existence of the function g is guaranteed by Theorem 2.10). If A is the matrix in Exercise 2.71 then $\det A = g(1,1) \ldots g(n,1)$.

2.73. Let k be a positive integer. If A is the $n \times n$ matrix $[(i,j)^k]$, $1 \leq i,j \leq n$, then $\det A = J_k(1) \ldots J_k(n)$. See Exercise 1.19 for the case $k = 1$. (Hint: use Exercise 2.40.)

2.74. Let k be a positive integer. If A is the $n \times n$ matrix $[c_k(i,j)]$, $1 \leq i,j \leq n$, and if $n \geq 2$ and $k \geq 2$, then $\det A = 0$.

Notes on Chapter 2

The sum $c(n,r)$ was introduced by S. Ramanujan [18], and thus it bears his name. The number on the right-hand side of the equation in Corollary 2.4 was called the von Sterneck number, and denoted by $\Phi(n,r)$, by C. A. Nicol and H. S. Vandiver [54]. They discussed the work of R. D. von Sterneck, giving the appropriate references, and gave new proofs of several of his results. The equality of $c(n,r)$ and $\Phi(n,r)$ was proved by O. Hölder [36], who proved Proposition 2.1 and gave the evaluation of $c(n,p^{\alpha})$ following Corollary 2.2. Another proof that $c(n,r) = \Phi(n,r)$ was given by E. Gagliardo [53].

Theorem 2.3 was proved by D. R. Anderson and T. M. Apostol [53]. Their generalized Ramanujan sums, which are the subject of that theorem, were studied further by T. M. Apostol [72]. The Brauer-Rademacher identity was stated as a problem by H. Rademacher [25], and a solution to the problem was given by A. Brauer [26]. Theorem 2.5 was proved by E. Cohen [60d], in a paper which contains a proof of Theorem 2.3 and a proof of the identity in Exercise 2.66. Among the papers written on the Brauer-Rademacher identity are those of E. Cohen [60a], [60k], M. V. Subbarao [65], A. C. Vasu [65] and P. Szüsz [67].

The Brauer-Rademacher identity is a special case of the very general identity derived in the paragraphs preceding Proposition 2.12. The general identity was obtained by E. Cohen [59b], and it contains also the result of P. Kesava Menon [65] stated in Exercise 1.7: the verification of this statement is the content of Exercise 2.40.

Theorem 2.6 was proved by E. Cohen [52], and he pointed out that it implies the orthagonality properties between Ramanujan sums discovered by R. D. Carmichael [32] and contained in Exercise 2.22. Theorem 2.8 was stated by E. Cohen [55b]. In this paper he introduced the notion of an even function (mod r), and proved Theorems 2.9 and 2.10. In a later paper, E. Cohen [58b] pointed out that the second formula for $\alpha(d)$ in Theorem 2.9 had been obtained earlier by K. G. Ramanathan [44]. E. Cohen continued his studies of even functions (mod r) in his papers [58b], [59b] and [59d] and in other papers, and he wrote an expository article [60f] on the subject. Many of the results that appear in the text after Theorem 2.10, and the results in a number of exercises, are from E. Cohen's papers.

The identity in Exercise 2.50 was discovered by P. J. McCarthy [62], after the special case at the end of the exercise was published by E. Cohen [59b].

The Ramanujan sums have been generalized in various directions. The generalization defined in Exercise 2.51 was made by E. Cohen [49], and the results in Exercises 2.51-2.58 were obtained by him in that paper and in others [55a], [56a]. Additional properties of $c_k(n,r)$ were verified by P. J. McCarthy [60c]. The (r,k)-even functions were defined and studied by P. J. McCarthy [60a], [60d], and the results in Exercises 2.59-2.61 are from those papers.

The sum $c^{(k)}(n,r)$ of Exercises 2.62 and 2.63 was defined by E. Cohen [59a], mentioned by him in another paper [60e], and in turn generalized by M. Suganamma [60]. A search through the bibliography will turn up other generalizations of the Ramanujan sums, and we shall mention two of them because they are off the path we have been trodding. The first is the extension of ideas surrounding the Ramanujan sums to a setting involving algebraic number fields. This was done by H. Rademacher [38], and the sums he defined are called Rademacher sums. They were rediscovered years later by G. J. Rieger [60]. The other extension, this time to a matrix setting, was made by K. G. Ramanathan and M. V. Subbarao [80].

The results in Exercises 2.67 and 2.68 are due to R. Sivaramakrishnan [79]. Those in Exercises 2.67-2.69 are new, but were suggested by a paper of D. Suryanarayana [78a].

Smith's determinant, given in Exercise 1.19, was evaluated first by H. J. S. Smith (see L. E. Dickson [D], Chapter V), who also gave the generalization in Exercise 2.73. The determinant in Exercise 2.70 was evaluated by T. M. Apostol [72], and those in Exercises 2.71, 2.72 and 2.74 by P. J. McCarthy [86]. Related results can be found in Exercises 3.30, 3.31, 4.33 and 4.34.

An exposition of the properties of the Ramanujan sums and related sums and functions was published by K. Nageswara Rao and R. Sivaramakrishnan [81].

Chapter 3
Counting Solutions of Congruences

In this chapter we shall use the results obtained in the preceding chapter to count solutions of certain linear and other congruences in s unknowns. By a solution of a congruence, with modulus r, we mean a solution (mod r), i.e., an ordered s-tuple of integers $\langle x_1, \ldots, x_s \rangle$ that satisfies the congruence, with two s-tuples $\langle x_1, \ldots, x_s \rangle$ and $\langle x_1', \ldots, x_s' \rangle$ that satisfy the congruence counted as the same solution if and only if $x_i \equiv x_i' \pmod{r}$ for $i = 1, \ldots, s$.

We shall count either all the solutions or all the solutions that are restricted in some way. For example, we might consider those solutions $\langle x_1, \ldots, x_s \rangle$ such that $(x_i, r) = 1$ for $i = 1, \ldots, s$.

We begin by counting the unrestricted solutions of the general linear congruence.

Proposition 3.1. The congruence

$$n \equiv a_1X_1 + \ldots + a_sX_s \pmod{r}$$

has a solution if and only if

$$d|n \text{ , where } d = (a_1, \ldots, a_s, r) \text{ .}$$

If it does have a solution, then it has dr^{s-1} solutions.

Proof. The condition that $d|n$ is certainly necessary for the congruence to have a solution.

On the other hand, suppose that $d|n$. We shall show, by induction on s , that the congruence has dr^{s-1} solutions.

Suppose that $s = 1$. The congruence

$$\frac{n}{d} \equiv \frac{a_1}{d} X_1 \pmod{\frac{r}{d}}$$

has a unique solution x_1 : hence $n \equiv a_1x_1 \pmod r$ has exactly d solutions, to wit, $x_1, x_1 + \frac{r}{d}, x_1 + 2\frac{r}{d}, \ldots, x_1 + (d-1)\frac{r}{d}$.

Now suppose that $s > 1$ and that the assertion is true for linear congruences with $s - 1$ unknowns. Let $e = (a_2, \ldots, a_s, r)$. Since $d = (a_1, e)|n$, the congruence $n \equiv a_1X_1 \pmod e$ has d solutions. Hence, in every complete residue system (mod r) there are $(r/e)d$ solutions of this congruence.

Let x_1 be a solution of $n \equiv a_1X_1 \pmod e$ and consider the congruence

$$n - a_1x_1 \equiv a_2X_2 + \ldots + a_sX_s \pmod r .$$

Since $e|n - a_1x_1$, it has er^{s-2} solutions. Therefore, the congruence with s unknowns has $(r/e)der^{s-2} = dr^{s-1}$ solutions. □

Now consider the congruence

$$(*) \qquad n \equiv X_1 + \ldots + X_s \pmod r .$$

We wish to count the solutions $\langle x_1, \ldots, x_s\rangle$ of this congruence for which the greatest common divisors $(x_i, r), i = 1, \ldots, s$, are restricted in various ways. (See Exercise 3.1.)

Let $N(n,r,s)$ be the number of solutions $\langle x_1, \ldots, x_s \rangle$ of (*) such that $(x_i, r) = 1$ for $i = 1, \ldots, s$.

Proposition 3.2. $N(\cdot,r,s)$ is an even function (mod r).

Proof. Let $n = n_1 n_2$, where $(n_2, r) = 1$: we claim that $N(n,r,s) = N(n_1, r, s)$. To each solution $\langle y_1, \ldots, y_s \rangle$ of

$$n_1 \equiv Y_1 + \ldots + Y_s \pmod{r}$$

there corresponds a solution of (*), to wit, $\langle n_2 y_1, \ldots, n_2 y_s \rangle$, and this is a one-one correspondence between the solutions of the two congruences. Furthermore, $(y_i, r) = 1$ for $i = 1, \ldots, s$ if and only if $(n_2 y_i, r) = 1$ for $i = 1, \ldots, s$. This proves the claim, and because of its truth and Exercise 3.2, it is enough to show that

$$N(n,r,s) = N((n,r),r,s)$$

whenever n and r are powers of the same prime p. We shall also use the obvious fact that $N(\cdot,r,s)$ is periodic (mod r).

Let $n = p^\alpha$ and $r = p^\beta$. If $\alpha \leq \beta$ then $(n,r) = p^\alpha = n$ and there is no proof required. Suppose $\alpha > \beta$. Then $p^\alpha + p^\beta = p^\beta m$, where $p \nmid m$, and

$$N(p^\alpha,p^\beta,s) = N(p^\alpha + p^\beta,p^\beta,s) = N(p^\beta m,p^\beta,s) = N(p^\beta,p^\beta,s) ,$$

which was to be proved. □

Thus,

$$N(n,r,s) = \sum_{d|r} \alpha(d)c(n,d) \quad \text{for all } n ,$$

and it remains to determine the Fourier coefficients $\alpha(d)$.

Theorem 3.3. For all n ,

$$N(n,r,s) = \frac{1}{r} \sum_{d|r} c(r/d,r)^s \, c(n,d) .$$

Proof. The proof will be by induction on s . The congruence $n \equiv X_1 \pmod r$ has one solution x_1 with $(x_1, r) = 1$ if $(n,r) = 1$, and no solutions otherwise: hence

$$N(n,r,1) = \begin{cases} 1 & \text{if } (n,r) = 1 \\ 0 & \text{if } (n,r) > 1 , \end{cases}$$

i.e., $N(n,r,1) = \delta((n,r))$. It was shown in Chapter 2 (following Proposition 2.11) that this is equal to

$$\frac{1}{r} \sum_{d|r} c(r/d,r) c(n,d) .$$

Now suppose that $s > 1$ and that

$$N(n,r,s-1) = \frac{1}{r} \sum_{d|r} c(r/d,r)^{s-1} \, c(n,d) \quad \text{for all } n .$$

Since it is certainly true that

$$N(n,r,s) = \sum_{n \equiv a+b \pmod r} N(a,r,1) N(b,r,s-1) ,$$

the assertion of the theorem follows from Proposition 2.11. □

For each divisor d of r,

$$c(r/d,r) = \frac{\phi(r)\mu(d)}{\phi(d)} .$$

Thus, for all n,

$$N(n,r,s) = \frac{\phi(r)^s}{r} \sum_{d|r} \frac{\mu(d)^s}{\phi(d)^s} c(n,d) .$$

If $r = p_1^{\alpha_1} \dots p_t^{\alpha_t}$ then

$$N(n,r,s) = \prod_{i=1}^{t} N(n,p_i^{\alpha_i},s) ,$$

and for a prime p and $\alpha \geq 1$,

$$\begin{aligned} N(n,p^\alpha,s) &= \frac{\phi(p^\alpha)^s}{p^\alpha} \sum_{j=0}^{\alpha} \frac{\mu(p^j)^s}{\phi(p^j)^s} c(n,p^j) \\ &= p^{\alpha(s-1)-s}(p-1)^s \left(1 + \frac{(-1)^s}{(p-1)^s} c(n,p)\right) \\ &= p^{\alpha(s-1)} \left(\frac{(p-1)^s + (-1)^s c(n,p)}{p^s}\right) . \end{aligned}$$

Since

$$c(n,p) = \begin{cases} \phi(p) = p - 1 & \text{if } p|n \\ \mu(p) = -1 & \text{if } p\nmid n , \end{cases}$$

we have

$$N(n,p^{\alpha},s) = \begin{cases} p^{\alpha(s-1)} \dfrac{(p-1)((p-1)^{s-1} - (-1)^{s-1})}{p^s} & \text{if } p \mid n \\ p^{\alpha(s-1)} \dfrac{(p-1)^s - (-1)^s}{p^s} & \text{if } p \nmid n . \end{cases}$$

Therefore, for all n ,

$$N(n,r,s) = r^{s-1} \prod_{p \mid (n,r)} \frac{(p-1)((p-1)^{s-1} - (-1)^{s-1})}{p^s} \prod_{\substack{p \mid r \\ p \nmid n}} \frac{(p-1)^s - (-1)^s}{p^s}$$

The <u>Nagell function</u> $\theta(\cdot,r)$ is defined by

$$\theta(n,r) = N(n,r,2) \quad \text{for all } n .$$

Thus,

$\theta(n,r)$ = the number of integers x such that $1 \leq x \leq r$ and $(x,r) = (n - x, r) = 1$

If p is a prime and $\alpha \geq 1$, then

$$\theta(n,p^{\alpha}) = \begin{cases} p^{\alpha-1}(p-1) & \text{if } p \mid n \\ p^{\alpha-1}(p-2) & \text{if } p \nmid n . \end{cases}$$

<u>Proposition 3.4</u>. For all n ,

$$\theta(n,r) = \phi(r) \sum_{\substack{d|r \\ (n,d)=1}} \frac{\mu(d)}{\phi(d)} .$$

Proof. Let γ be the core function defined in Exercise 1.14. If $d|r$ then $\mu(d) = 0$ unless $d|\gamma(r)$: hence

$$\theta(n,r) = \frac{\phi(r)^2}{r} \sum_{d|\gamma(r)} \frac{\mu(d)^2}{\phi(d)^2} c(n,d) .$$

If, for every divisor d of $\gamma(r)$, we let $\alpha(d) = \mu(d)/\phi(d)$ in Proposition 2.12, the result is

$$\sum_{d|\gamma(r)} \frac{\mu(d)^2}{\phi(d)^2} c(n,d)$$

$$= \sum_{\substack{d|\gamma(r) \\ (n,d)=1}} \frac{d}{\phi(d)} \sum_{D|\frac{\gamma(r)}{d}} \frac{\mu(Dd)}{\phi(Dd)} \mu(D) .$$

However, $(d,D) = 1$ whenever D divides $\gamma(r)/d$, and so the inner sum is equal to

$$\frac{\mu(d)}{\phi(d)} \sum_{D|\frac{\gamma(r)}{d}} \frac{|\mu(D)|}{\phi(d)} = \frac{\mu(d)}{\phi(d)} \frac{\gamma(r)}{d} \frac{1}{\phi(\gamma(r)/d)} ,$$

where we have used the final identity in Chapter 2. Since $\phi(d)\phi(\gamma(r)/d) = \phi(\gamma(r))$ we have, therefore

$$\theta(n,r) = \frac{\phi(r)^2}{r} \frac{\gamma(r)}{\phi(\gamma(r))} \sum_{\substack{d|\gamma(r) \\ (n,d)=1}} \frac{\mu(d)}{\phi(d)} .$$

In the sum we can replace $\gamma(r)$ by r. Furthermore, since r and $\gamma(r)$ have the same prime divisors,

$$\frac{\gamma(r)}{\phi(\gamma(r))} = \frac{r}{\phi(r)} .$$

Thus, the formula for $\theta(n,r)$ is that of the proposition. □

The most general problem of the type that was solved in Theorem 3.3 is the following one. Let r be a positive integer and for $i = 1, \ldots, s$ let $T_i(r)$ be a nonempty subset of $\{1, \ldots, r\}$. The problem is to determine the number $M(n,r,s)$ of solutions $\langle x_1, \ldots, x_s\rangle$ of (*) for which $x_i \in T_i(r)$ for $i = 1, \ldots, s$. We shall show that if the sets $T_i(r)$ have a certain property then the problem has a very neat solution. And we shall show how to produce, in a systematic fashion, sets $T_i(r)$ with that property.

Theorem 3.5. For $i = 1, \ldots, s$ let

$$g_i(n,r) = \sum_{x \in T_i(r)} e(nx,r) \quad \text{for all } n .$$

If each function $g_i(\cdot,r)$ is an even function (mod r) then

$$M(n,r,s) = \frac{1}{r} \sum_{d|r} \left(\prod_{i=1}^{s} g_i(r/d\ ,r) \right) c(n,d) .$$

Before proving the theorem we shall show how to produce sets $T_i(r)$ for which the hypothesis holds.

Proposition 3.6. Let $D(r)$ be a nonempty set of divisors of the positive integer r. Let

$$T(r) = \{x : 1 \le x \le r \text{ and } (x,r) \in D(r)\} .$$

If

$$g(n,r) = \sum_{x \in T(r)} e(nx,r) \quad \text{for all } n ,$$

then $g(\cdot,r)$ is an even function (mod r): in fact, for all n,

$$g(n,r) = \sum_{d \in D(r)} c(n,r/d) .$$

Proof. By Exercise 3.6,

$$g(n,r) = \sum_{d \in D(r)} \sum_{\substack{1 \le x \le r/d \\ (x,r/d)=1}} e(nx,r/d) ,$$

and the inner sum is equal to $c(n,r/d)$. □

Example. Let $D(r) = \{1\}$: then $x \in T(r)$ if and only if $1 \le x \le r$ and $(x,r) = 1$. Thus, $M(n,r,s) = N(n,r,s)$ and $g(n,r) = c(n,r)$, and Theorem 3.5 yields Theorem 3.3 as a special case.

Proof of Theorem 3.5. If we set $M = M(n,r,s)$ then

$$M = \sum \prod_{i=1}^{s} h_i(x_i)$$

where

$$h_i(x_i) = \begin{cases} 1 & \text{if } x_i \in T_i(r) \\ 0 & \text{otherwise}, \end{cases}$$

and where the sum $\sum$ is over all solutions of (*). Since

$$h_i(x_i) = \frac{1}{r} \sum{}' \sum_{q_i=1}^{r} e((x_i - y)q_i, r) ,$$

where the sum $\sum'$ is over all $y \in T_i(r)$,

$$M = \frac{1}{r^s} \sum \prod_{i=1}^{s} \sum{}' \sum_{q_i=1}^{r} e((x_i - y)q_i, r)$$

$$= \frac{1}{r^s} \sum{}'' \sum \prod_{i=1}^{s} \sum{}' e(x_i q_i, r) e(-yq_i, r) ,$$

where the sum $\sum''$ is over all s-tuples $\langle q_1, \ldots, q_s \rangle$ of integers from the set $\{1, \ldots, r\}$. Thus,

$$M = \frac{1}{r^s} \sum{}'' \sum \prod_{i=1}^{s} e(x_i q_i, r) \left(\sum{}' e(-yq_i, r) \right) ,$$

and the sum in the brackets is equal to $g_i(q_i,r)$. Thus,

$$M = \frac{1}{r^s}\sum{}''\left(\prod_{i=1}^{s} g_i(q_i,r)\right)\sum\prod_{i=1}^{s} e(x_i q_i,r) .$$

By Exercise 3.7,

$$\sum\prod_{i=1}^{s} e(x_i q_i,r) = \begin{cases} r^{s-1}e(nq,r) & \text{if } q_1=\dots=q_s=q \\ 0 & \text{otherwise} . \end{cases}$$

Hence,

$$M = \frac{1}{r}\sum_{q=1}^{r}\left(\prod_{i=1}^{s} g_i(q,r)\right) e(nq,r)$$

$$= \frac{1}{r}\sum_{d|r}\sum_{\substack{1\le u\le r/d \\ (u,r/d)=1}}\left(\prod_{i=1}^{s} g_i(ud,r)\right) e(nu,r/d) ,$$

and $g_i(ud,r) = g_i(d,r)$ since $(ud,r) = d = (d,r)$, and $g_i(\cdot,r)$ is an even function (mod r) by hypothesis. Therefore,

$$M = \frac{1}{r}\sum_{d|r}\left(\prod_{i=1}^{s} g_i(d,r)\right) c(n,r/d)$$

$$= \frac{1}{r}\sum_{d|r}\left(\prod_{i=1}^{s} g_i(r/d,r)\right) c(n,d) . \ \square$$

The advantage of this proof is that it makes use of no property of even functions (mod r) other than the defining property. We shall give three more examples of the use of Theorem 3.5. Other examples are contained in exercises.

<u>Example</u>. Let $D(r)$ be the set of kth power divisors of r : then

$$T(r) = \{x : 1 \le x \le r \text{ and } (x,r) \text{ is a kth power}\}$$

and

$$g_k(n,r) = g(n,r) = \sum_{d^k|r} c(n,r/d^k) \quad \text{for all } n$$

($g_2(n,r) = B(n,r)$, defined in Exercise 2.64). Therefore, if

$P_k(n,r,s)$ = the number of solutions $\langle x_1, \ldots, x_s \rangle$ of (*) such that (x_i, r) is a kth power for $i = 1, \ldots, s$,

then

$$P_k(n,r,s) = \frac{1}{r} \sum_{d|r} g_k(r/d,r)^s \, c(n,d) \; .$$

<u>Example</u>. Let $D(r)$ be the set of all divisors of r that are k-free, i.e., divisible by no kth power greater than one. Then

$$T(r) = \{x : 1 \le x \le r \text{ and } (x,r) \text{ is k-free}\}$$

and

$$h_k(n,r) = g(n,r) = \sum_{\substack{d|r \\ d \text{ k-free}}} c(n,r/d) .$$

Therefore, if

$Q_k(n,r,s)$ = the number of solutions $\langle x_1,\ldots,x_s\rangle$ of (*) such that (x_i,r) is k-free for $i = 1, \ldots, s$,

then

$$Q_k(n,r,s) = \frac{1}{r}\sum_{d|r} h_k(r/d,r)^s \, c(n,d) .$$

<u>Example</u>. Let $N_k(n,r,s)$ be the number of solutions $\langle x_1, \ldots, x_s\rangle$ of the congruence

$$n \equiv x_1 + \ldots + x_s \pmod{r^k}$$

such that $(x_i, r^k)_k = 1$ for $i = 1, \ldots, s$ (see Exercise 1.29). The restriction is simply that (x_i, r^k) be k-free for $i = 1, \ldots, s$: hence $N_k(n,r,s) = Q_k(n,r^k,s)$. By Proposition 3.6,

$$h_k(n,r^k) = \sum_{\substack{d|r^k \\ (d,r^k)_k = 1}} c(n,r^k/d) ,$$

and this is equal to $c_k(n,r)$ (see Exercises 2.51 and 2.54). Therefore,

$$N_k(n,r,s) = \frac{1}{r^k}\sum_{d|r^k} c_k(r^k/d,r)^s \, c(n,d) .$$

Another evaluation of $N_k(n,r,s)$ is given in Exercise 3.13.

Other restrictions can be placed on solutions of (*). For example, we can count the solutions $\langle x_1, \ldots, x_s\rangle$ of (*) for which $(x_1, \ldots, x_s, r) = 1$. In fact, we can just as easily solve a more general problem.

Theorem 3.7. Let $a_1, \ldots, a_s$ be integers such that $(a_1, \ldots, a_s, r) = 1$. If $N'(n,r,s)$ is the number of solutions $\langle x_1, \ldots, x_s\rangle$ of the congruence

$$n \equiv a_1X_1 + \ldots + a_sX_s \pmod r$$

such that $(x_1, \ldots, x_s, r) = 1$, then

$$N'(n,r,s) = \sum_{d \mid (n,r)} (r/d)^{s-1}\mu(d) .$$

Proof. Let $(n,r) = p_1^{\alpha_1} \ldots p_t^{\alpha_t}$, and for $i = 1, \ldots, t$ let A_i be the set of solutions $\langle x_1, \ldots, x_s\rangle$ of the congruence such that $p_i \mid (x_1, \ldots, x_s, r)$. Then $N'(n,r,s)$ is the number of solutions of the congruence that are not in $A_1 \cup \ldots \cup A_t$. If $1 \le i_1 < \ldots < i_j \le t$ then $\#(A_{i_1} \cap \ldots \cap A_{i_j})$ is the number of unrestricted solutions of the congruence

$$\frac{n}{p_{i_1} \cdots p_{i_j}} \equiv a_1Y_1 + \ldots + a_sY_s \left(\bmod \frac{r}{p_{i_1} \cdots p_{i_j}}\right) .$$

Since $(a_1, \ldots, a_s, r/p_{i_1} \cdots p_{i_j}) = 1$ this number is $(r/p_{i_1} \cdots p_{i_j})^{s-1}$ by Proposition 3.1. Thus, by the Inclusion-Exclusion Principle,

$$N'(n,r,s) = r^{s-1} + \sum_{j=1}^{t} (-1)^j \sum_{1 \le i_1 < \ldots < i_j \le t} \left(\frac{r}{p_{i_1} \cdots p_{i_j}}\right)^{s-1}$$

$$= \sum_{d|(n,r)} (r/d)^{s-1} \mu(d). \square$$

Note that $N'(n,r,s)$ is independent of the coefficients of the congruence $a_1, \ldots, a_s$. The only way in which they enter into the result is in the hypothesis that $(a_1, \ldots, a_s, r) = 1$. Thus, if $a = (a_1, \ldots, a_s)$ and $(a,r) = 1$ then $N'(n,r,s)$ is the number of solutions $\langle x_1, \ldots, x_s \rangle$ of

$$n \equiv a(x_1 + \ldots + x_s) \pmod{r}$$

such that $(x_1, \ldots, x_s, r) = 1$.

A congruence of the form

$$(**) \qquad n \equiv a_1 X_1 Y_1 + \ldots + a_s X_s Y_s \pmod{r}$$

is called a <u>semi-linear congruence</u>. We shall assume that $(a_i, r) = 1$ for $i = 1, \ldots, s$. A solution of (**) is a 2s-tuple of integers $\langle x_1, \ldots, x_s, y_1, \ldots, y_s \rangle$. Let $S(n,r,s)$ denote the number of solutions of (**).

<u>Theorem 3.8.</u> For all n,

$$S(n,r,s) = r^{2s-1} \sum_{d|r} \frac{1}{d^s} c(n,d).$$

Note that, assuming the truth of the theorem,

$$S(n,r,s) = \tau_{s-1}(n,r) = r^{s-1} \sum_{d|(n,r)} d\, J_s(r/d)$$

(see Exercise 2.37). In particular, if $(n,r) = 1$,

$$S(n,r,s) = r^{s-1} J_s(r) .$$

Proof. The proof will be by induction on s. For $s = 1$ we must show that

$$s(n,r,1) = \sum_{d|(n,r)} d\phi(r/d) .$$

By Exercise 3.2, it is enough to do this when $r = p^\alpha$, where p is a prime and $\alpha \geq 1$.

Out task is to count the solutions of

$$n \equiv a X Y \pmod{p^\alpha} , \ p \nmid a .$$

Let $(n,p^\alpha) = p^\beta$. Choose j and b such that $0 \leq j \leq \beta$, $1 \leq b \leq p^{\alpha-j}$ and $p \nmid b$: for each choice of j there are $\phi(p^{\alpha-j})$ choices for b. Then the congruence

$$n \equiv a p^j b Y \pmod{p^\alpha}$$

has p^j solutions, and if y is one of its solutions then $\langle p^j b, y\rangle$ is a solution of the semi-linear congruence. Thus, for each choice for j, there are at least $\phi(p^{\alpha-j})p^j$ solutions of the semi-linear congruence.

In fact, every solution of the semi-linear congruence arises in this way: if $\langle x,y\rangle$ is a solution and if $x = p^j b$ where $1 \leq x \leq p^\alpha$ and

and $p \nmid b$, then $0 \le j \le \beta$ and $1 \le b \le p^{\alpha-j}$. Therefore,

$$S(n,p^{\alpha},1) = \sum_{j=0}^{\beta} \phi(p^{\alpha-j})p^j ,$$

which was to be shown.

Now assume that $s > 1$ and that the formula of the theorem holds when s is replaced by $s-1$. Then

$$S(n,r,s) = \sum_{n\equiv a+b \pmod r} S(a,r,1)S(b,r,s-1)$$

$$= \sum_{n\equiv a+b \pmod r} \tau_0(a,r)\tau_{s-2}(b,r)$$

$$= \tau_{s-1}(n,r) \qquad \text{by Exercise 2.38}$$

$$= r^{2s-1} \sum_{d|(n,r)} \frac{1}{d^s} c(n,d). \quad \square$$

At this point there are two ways in which we can go. We can consider several linear or semi-linear congruences simultaneously, or we can consider other nonlinear congruences. In order to count the solutions of simultaneous congruences we require another generalization of the Ramanujan sums.

For a positive integer r and integers $n_1, \ldots, n_t$ let

$$c(n_1, \ldots, n_t, r) = \sum_{(x_1,\ldots,x_t,r)=1} e(n_1x_1 + \ldots + n_tx_t,r) .$$

The sum is over a reduced t-residue system (mod r) (see Exercise 1.41). Let $n = (n_1, \ldots, n_t)$, and if $d = (n,r)$ let $m_i = n_i/d$ for $i = 1, \ldots, t$ and $m = (m_1, \ldots, m_t) = n/d$. By Exercise 1.41,

$$c(n_1, \ldots, n_t, r) = \frac{J_t(r)}{J_t(r/d)} \sum_{(y_1,\ldots,y_t,r/d)=1} e(m_1y_1 + \ldots + m_ty_t, r/d) ,$$

and by the remark that follows the proof of Theorem 3.7, this is equal to

$$\frac{J_t(r)}{J_t(r/d)} \sum_{(y_1,\ldots,y_t,r/d)=1} e(m(y_1 + \ldots + y_t), r/d)$$

$$= \sum_{(x_1,\ldots,x_t,r)=1} e(n(x_1 + \ldots + x_t), r) = c^{(t)}(n,r) ,$$

where $c^{(t)}(n,r)$ is the sum defined in Exercise 2.62. Therefore, as shown in that exercise,

$$c(n_1, \ldots, n_t, r) = \sum_{d|(n_1,\ldots,n_t,r)} d^t\mu(r/d) .$$

A function f of t-integer variables is called a <u>totally even function (mod r)</u> if there is an even function (mod r), say F, such that

$$f(n_1, \ldots, n_t) = F((n_1, \ldots, n_t)) \quad \text{for all } n_1, \ldots, n_t :$$

F is called the <u>associated</u> even function. $c(\cdot, \ldots, \cdot, r)$ is a totally even function (mod r), with associated even function $c^{(t)}(\cdot,r)$, as is any function f such that

$$f(n_1, \ldots, n_t) = \sum_{d|r} \alpha(d) c(n_1, \ldots, n_t, d)$$

for all $n_1, \ldots, n_t$.

For $j = 1, \ldots, s$ let $T_j(r)$ be a nonempty set of order t-tuples of integers from the set $\{1,\ldots,r\}$. Let $M(n_1, \ldots, n_t, r, s)$ be the number of solutions

$$\langle x_{11}, \ldots, x_{1s}, x_{21}, \ldots, x_{2s}, \ldots, x_{t1}, \ldots, x_{ts}\rangle$$

of the system of linear congruences

$$(\#) \qquad n_i \equiv X_{i1} + \ldots + X_{is} \pmod{r}, \; i = 1, \ldots, t ,$$

such that $\langle x_{ij}, \ldots, x_{tj}\rangle \in T_j(r)$ for $j = 1, \ldots, s$. Also for $j = 1, \ldots, s$ let g_j be the function defined by

$$g_j(n_1, \ldots, n_t) = \sum_{\langle x_1,\ldots,x_t\rangle \in T_j(r)} e(n_1x_1 + \ldots + n_tx_t, r) .$$

We have the following generalization of Theorem 3.5.

<u>Theorem 3.9</u>. If g_j is a totally even function (mod r), with associated even function G_j , for $j = 1, \ldots, s$, then

$$M(n_1, \ldots, n_t, r) = \frac{1}{r^t} \sum_{d|r} \left(\prod_{j=1}^{s} G_j(r/d) \right) c(n_1, \ldots, n_t, d) .$$

The proof is similar to that of Theorem 3.5: only an outline will be given, with the details left as an exercise.

If $M = M(n_1, \ldots, n_t, r)$ then

$$M = \sum_1 \cdots \sum_t \prod_{j=1}^{s} h_j(x_{1j}, \ldots, x_{tj}) ,$$

where the sum $\sum_i$ is over all solutions of the ith congruence and

$$h_j(x_{1j}, \ldots, x_{tj}) = \begin{cases} 1 & \text{if } \langle x_{1j}, \ldots, x_{tj}\rangle \in T_j(r) \\ 0 & \text{otherwise} \end{cases}$$

$$= \frac{1}{r^t} \sum_{(j)} \prod_{j=1}^{t} \sum_{q_{ij}=1}^{r} e((x_{ij} - y_i)q_{ij}, r) ,$$

where the sum $\sum_{(j)}$ is over all $\langle y_1, \ldots, y_t\rangle \in T_j(r)$. After some manipulation we obtain

$$M = \frac{1}{r^{ts}} \sum{}' \left(\prod_{j=1}^{s} g_j(q_{1j}, \ldots, q_{tj}, r) \right) \prod_{i=1}^{t} \sum_i \sum_{j=1}^{s} e(x_{ij}q_{ij}, r) ,$$

where the sum $\sum'$ is over all ordered ts-tuples of integers from $\{1, \ldots, r\}$. By Exercise 3.7, this is equal to

$$\frac{1}{r^t} \sum{}'' \left(\prod_{j=1}^{s} g_j(q_1, \ldots, q_t, r) \right) \prod_{i=1}^{t} e(n_i q_i, r) ,$$

where the sum $\sum''$ is over all ordered t-tuples $\langle q_1, \ldots, q_t\rangle$ of integers from $\{1, \ldots, r\}$. By Exercise 3.22,

$$M = \frac{1}{r^t}\sum_{d|r}\ \sum_{(u_1,\ldots,u_t,r/d)=1}\left(\prod_{j=1}^{s} g_j(u_1 d, \ldots, u_t d, r)\right)$$

$$\cdot\, e(n_1u_1 + \ldots + n_tu_t, r/d)\ .$$

Now $g_j(u_1d, \ldots, u_td, r/d) = g_j(d, \ldots, d, r) = G_j(d)$, and from the definition of $c(n_1, \ldots, n_t, r)$,

$$M = \frac{1}{r^t}\sum_{d|r}\left(\prod_{j=1}^{s} G_j(d)\right) c(n_1, \ldots, n_t, r/d)\ .$$

This completes the outline of the proof of Theorem 3.9.

Let $D(r)$ be a nonempty set of divisors of r and let

$$T(r) = \{\langle x_1, \ldots, x_t\rangle : 1 \le x_i \le r \text{ for } i = 1, \ldots, t \text{ and}$$

$$(x_1, \ldots, x_t, r) \in D(r)\}\ .$$

By Exercise 3.22,

$$g(n_1, \ldots, n_t, r) = \sum_{\langle x_1,\ldots,x_t\rangle \in T(r)} e(n_1x_1 + \ldots + n_tx_t, r)$$

$$= \sum_{d\in D(r)} c(n_1, \ldots, n_t, r/d)\ .$$

Thus, g is a totally even function (mod r), and if G is the associated even function, then for all n,

$$G(n) = \sum_{d \in D(r)} c^{(t)}(n, r/d)$$

Example. If $N(n_1, \ldots, n_t, r, s)$ is the number of solutions $\langle x_{ij} \rangle$ of (#) with $(x_{ij}, \ldots, x_{tj}, r) = 1$ for $j = 1, \ldots, s$, then

$$N(n_1, \ldots, n_j, r, s) = \frac{1}{r^t} \sum_{d|r} c^{(t)}(r/d, r)^s c(n_1, \ldots, n_t, d) .$$

In particular, with $t = 2$, the number of solutions $\langle x_1, \ldots, x_s, y_1, \ldots, y_s \rangle$ of the pair of congruences

$$(\#\#) \qquad \begin{aligned} m &\equiv X_1 + \ldots + X_s \pmod{r} \\ n &\equiv Y_1 + \ldots + Y_s \pmod{r} \end{aligned}$$

such that $(x_i, y_i, r) = 1$ for $i = 1, \ldots, s$ is

$$N(m,n,r,s) = \frac{1}{r^2} \sum_{d|r} c^{(2)}(r/d, r)^s \; c(m,n,d)$$

$$= \frac{J_2(r)^s}{r^2} \sum_{d|r} \frac{\mu(d)^s}{J_2(d)^s} \; c(m,n,d) .$$

A generalization of Nagell's function can be defined by setting

$$\theta(m,n,r) = N(m,n,r,2) = \frac{J_2(r)^2}{r^2} \sum_{d|r} \frac{|\mu(d)|}{J_2(d)^2} \; c(m,n,d)$$

for all m and n. For a prime p and $\alpha \geq 1$,

$$\theta(m,n,p^\alpha) = \frac{J_2(p^\alpha)^2}{p^{2\alpha}} \left(1 + \frac{c(m,n,p)}{J_2(p)^2}\right) ,$$

and since

$$c(m,n,p) = \begin{cases} p^2 - 1 & \text{if } p \mid (m,n) \\ -1 & \text{if } p \nmid (m,n) , \end{cases}$$

we have

$$\theta(m,n,p^\alpha) = \begin{cases} p^{2(\alpha-1)}(p^2-1) & \text{if } p \mid (m,n) \\ p^{2(\alpha-1)}(p^2-2) & \text{if } p \nmid (m,n) . \end{cases}$$

Therefore, $\theta(m,n,r) \neq 0$ for all m and n.

Exercises

3.1. The solutions of (*) are in one-one correspondence with the solutions of the congruence

$$(*') \qquad n \equiv a_1X_1 + \dots + a_sX_s \pmod r , \ (a_1 \dots a_s, r) = 1 .$$

In fact, there is a correspondence such that if the solution $\langle x_1, \dots, x_s\rangle$ of (*) corresponds to the solution $\langle x_1', \dots, x_s'\rangle$ of (*') then $(x_i,r) = (x_i',r)$ for $i = 1, \dots, s$. Thus, the number of solutions of (*) with each of the restrictions we shall consider is equal to the number of restrictions of (*') with the same restriction.

3.2. Let $M(n,r,s)$ be the number of solutions (mod r) of a congruence

$$n \equiv P(X_1, \ldots, X_s) \pmod{r} ,$$

where P is a polynomial with integer coefficients. If $(r,r') = 1$ then $M(n,rr',s) = M(n,r,s)M(n,r',s)$. (Hint: use the Chinese remainder theorem.)

3.3. For all n ,

$$N(n,r,s+1) = \sum_{(x,r)=1} N(n+x, r,s)$$

and $N(n,r,s) = d^{s-1}N(n,r/d,s)$, where d is the largest square divisor of r .

3.4. If r is a positive integer and n is an integer, then $n \equiv X + Y \pmod{r}$ has a solution $\langle x,y \rangle$ with $(x,r) = (y,r) = 1$ if and only if either r is odd or both r and n are even.

3.5. If r is an even positive integer and n is odd, then $N(n,r,3) > 0$.

3.6. Let $D(r)$ and $T(r)$ be as in Proposition 3.6. If $d \in D(r)$ let

$$S_d = \{xd : 1 \le x \le r/d \text{ and } (x,r/d) = 1\} .$$

If $d \neq e$ then $S_d \cap S_e$ is empty and

$$\bigcup_{d \in D(r)} S_d = T(r) .$$

3.7. Let r be a positive integer and let $1 \le q_i \le r$ for $i = 1, \ldots, s$. For every n,

$$\sum_{n \equiv x_1 + \cdots + x_s \pmod r} e(x_1 q_1, r) \ldots e(x_s q_s, r) = \begin{cases} r^{s-1} e(nq, r) & \text{if } q_1 = \cdots = q_s = q \\ 0 & \text{otherwise} . \end{cases}$$

(See Exercise 2.18.)

3.8. Let $d_1, \ldots, d_s$ be divisors of r. The number of solutions $\langle x_1, \ldots, x_s \rangle$ of (*) such that $(x_i, r) = d_i$ for $i = 1, \ldots, s$ is

$$\frac{1}{r} \sum_{d|r} c(r/d, r/d_1) \ldots c(r/d, r/d_s) c(n, d) .$$

3.9. Let k and q be integers with $0 < q < k$ and let $S_{k,q}$ be the set of (k,q)-integers (see Exercise 1.92). If

$$D_{k,q}(n, r) = \sum_{(x,r) \in S_{k,q}} e(nx, r)$$

then for all n and r,

$$D_{k,q}(n, r) = \sum_{d|(n,r)} d\, \lambda_{k,q}(r/d) = \sum_{\substack{d|r \\ d \in S_{k,q}}} c(n, r/d) .$$

If $P_{k,q}(n, r, s)$ is the number of solutions $\langle x_1, \ldots, x_s \rangle$ of (*) such that $(x_i, r) \in S_{k,q}$ for $i = 1, \ldots s$, then

$$P_{k,q}(n,r,s) = \frac{1}{r}\sum_{d|r} D_{k,q}(r/d,r)^s\, c(n,d)$$

Note that $P_{k,1}(n,r,s) = P_k(n,r,s)$.

3.10. The number of solutions $\langle x_1, \ldots, x_s, y_1, \ldots, y_t\rangle$ of

$$n \equiv X_1 + \ldots + X_s + Y_1 + \ldots + Y_t \pmod r$$

such that $(x_i,r) = 1$ for $i = 1, \ldots, s$ and (y_i,r) is a kth power for $i = 1, \ldots, t$ is

$$\frac{1}{r}\sum_{d|r} c(r/d,r)^s\, g_k(r/d,r)^t\, c(n,d) .$$

3.11. Consider the arithmetical function ν_2 , defined in Exercise 1.89:

$$\nu_2(n) = \begin{cases} 1 & \text{if } n \text{ is a square} \\ 0 & \text{otherwise} . \end{cases}$$

If r is a positive integer then $\nu_2((\cdot,r))$ is an even function (mod r) and

$$\nu_2((n,r)) = \frac{1}{r}\sum_{d|r} \lambda(d)\beta(r/d)c(n,d) \quad \text{for all } n .$$

In particular,

$$\nu_2(r) = \frac{1}{r}\sum_{d|r} \lambda(d)\beta(r/d)\phi(d) \quad \text{for all } r .$$

3.12. For all n,

$$P_2(n,r,s) = \frac{1}{r}\sum_{d|r} (\lambda(d)\beta(r/d))^s c(n,d) .$$

(Hint: use Proposition 2.11.)

3.13. For all n,

$$N_k(n,r,s) = \frac{1}{r^k}\sum_{d|r} c_k((r/d)^k,r)^s\ c_k(n,d) .$$

3.14. Let $\theta_k(n,r) = N_k(n,r,2)$. For all n and all primes p and $\alpha \geq 1$,

$$\theta_k(n,p^\alpha) = \begin{cases} p^{k(\alpha-1)}(p^k-1) & \text{if } p^k|n \\ p^{k(\alpha-1)}(p^k-2) & \text{if } p^k\nmid n \end{cases}$$

Thus, $\theta_k(n,r) = 0$ if and only if $k = 1$, r is even and n is odd.

3.15. Let r be a positive integer and for all n let

$R(n,r)$ = the number of integers x such that $1 \leq x \leq r$, $(x,r) = 1$ and $(n - x, r)$ is a square.

Then

$$R(n,r) = \frac{\phi(r)}{r}\sum_{d|r} \frac{\mu(d)\lambda(d)}{\phi(d)}\beta(r/d)c(n,d)$$

and

$$R(n,r) = \phi(r) \sum_{\substack{d|r \\ (n,d)=1}} \frac{\lambda(d)}{\phi(d)} .$$

3.16. $N'(\cdot,r,s)$ is an even function (mod r). For all n,

$$N'(n,r,s) = \frac{1}{r} \sum_{d|r} c_s((r/d)^s,r)c(n,d) ,$$

where $c_s(n,r)$ is the generalized Ramanujan sum defined in Exercise 2.51.

3.17. For all n, and $s > 1$,

$$N'(n,r,s) = \left(\frac{r}{(n,r)}\right)^{s-1} J_{s-1}((n,r)) .$$

3.18. For a positive integer r, define the arithmetical function $\theta'(\cdot,r)$ by

$\theta'(n,r)$ = the number of integers x such that $1 \leq x \leq r$ and $(x, n-x, r) = 1$.

For all n,

$$\theta'(n,r) = \phi(r) \sum_{\substack{d|r \\ (n,d)=1}} \frac{|\mu(d)|}{\phi(d)} .$$

3.19. If r is a positive integer then $\theta'(n,r) \neq 0$ for all n.

3.20. Let r and s be positive integers. Then

$$c(n,r)^s = \sum_{d|r} N(r/d,r,s)c(n,d) \quad \text{for all } n .$$

In particular,

$$c(n,r)^2 = \sum_{d|r} \theta(r/d,r)c(n,d) \quad \text{for all } n ,$$

$$\sum_{d|r} N(r/d,r,s)\phi(d) = \phi(r)^s$$

and

$$\sum_{d|r} N(r/d,r,s)\mu(d) = \mu(r)^s .$$

3.21. This exercise is an alternate proof of Theorem 3.8.

(a) For every divisor d of (n,r) there are $r^{s-1} d J_s(r/d)$ solutions of (**) of the form $\langle b_1 d, \ldots, b_s d ; y_1, \ldots, y_s \rangle$ where $1 \le b_i \le r/d$ for $i = 1, \ldots, s$ and $(b_1, \ldots, b_s, r/d) = 1$.

(b) Every solution of (**) arises in this way.

3.22. (Compare with Exercise 3.6.) Let $D(r)$ be a nonempty set of divisors of r and let

$$T(r) = \{\langle x_1, \ldots, x_t \rangle : 1 \le x_i \le r \text{ for } i = 1, \ldots, t \text{ and } (x_1, \ldots, x_t, r) \in D(r)\} .$$

If $d \in D(r)$ let

$$S_d = \{\langle u_1 d, \ldots, u_t d \rangle : 1 \le u_i \le r/d \quad \text{for } i = 1, \ldots, t \text{ and } (u_1, \ldots, u_t, r/d) = 1\} .$$

If $d \neq e$ then $S_d \cap S_e$ is empty, and

$$\bigcup_{d \in D(r)} S_d = T(r) .$$

When $D(r) = \{1\}$ this result can be used to show that for all r,

$$J_t(r) = \sum_{d|r} d^t \mu(r/d) .$$

3.23. For all m and n,

$$c(m,n,r) = \sum_{\substack{d|m \\ e|n \\ [d,e]=r}} c(m,d)c(n,e) .$$

3.24. If $d, e|r$ then

$$\sum_{\substack{m \equiv a+b \pmod r \\ n \equiv a'+b' \pmod r}} c(a,a',d)c(b,b',e)$$

$$= \begin{cases} r^2 \, c(m,n,d) & \text{if } d = e \\ 0 & \text{if } d \neq e . \end{cases}$$

3.25. The number of unrestricted solutions of (#) is $r^{t(s-1)}$. If $N'(n_1, \ldots, n_t, r, s)$ is the number of solutions $\langle x_{ij} \rangle$ of (#) such that

$$(x_{11}, \ldots, x_{1s}, x_{21}, \ldots, x_{2s}, \ldots, x_{t1}, \ldots, x_{ts}, r) = 1 ,$$

then

$$N'(n_1, \ldots, n_t, r, s) = \sum_{d|(n_1,\ldots,n_t,r)} (r/d)^{t(s-1)} \mu(d) .$$

3.26. For all m and n,

$$N'(m,n,r,s) = \frac{1}{r^2} \sum_{d|r} c^{(2s)}(r/d,r)c(m,n,d)$$

$$= \frac{J_{2s}(r)}{r^2} \sum_{d|r} \frac{\mu(d)}{J_{2s}(d)} c(m,n,d) .$$

3.27. The number of solutions

$\langle x_1, \ldots, x_s, x_1', \ldots, x_t', y_1, \ldots, y_s, y_1', \ldots, y_t' \rangle$ of the congruences

$$m \equiv X_1 + \ldots + X_s + X_1' + \ldots + X_t' \pmod r$$

$$n \equiv Y_1 + \ldots + Y_s + Y_1' + \ldots + Y_t' \pmod r$$

such that

$$(x_i, y_i, r) = 1 \quad \text{for } i = 1, \ldots, s$$

and

$$(x_1', \ldots, x_t', y_1', \ldots, y_t', r) = 1$$

is

$$\frac{J_2(r)^s J_{2t}(r)}{r^2} \sum_{d|r} \frac{\mu(d)^{s+1}}{J_2(d)^s J_{2t}(d)} \quad c(m,n,d)$$

3.28. For all m and n ,

$$\theta(m,n,r) = J_2(r) \sum_{\substack{d|r \\ (m,n,d)=1}} \frac{\mu(d)}{J_2(d)}$$

3.29. If A is the $n \times n$ matrix $[N(i,j,s)]$, $1 \le i,j \le n$, then

$$\det A = \begin{cases} 1 & \text{if } n = 1, \text{ or } n = 2 \text{ and } s \text{ is even, or } n = 3 \\ -1 & \text{if } n = 2 \text{ and } s \text{ is odd} \\ 0 & \text{if } n \ge 4. \end{cases}$$

3.30. If B is the $n \times n$ matrix $[N'(i,j,s)]$, $1 \le i,j \le n$, then $\det B = \det A$, where A is the matrix in Exercise 3.29.

3.31. If A is the $n \times n$ matrix $[S(i,j,s)]$, $1 \le i,j \le n$, then $\det A = (n!)^s$.

Notes on Chapter 3

The formula for $N(n,r,s)$ derived in the paragraphs following the proof of Theorem 3.3 was stated as a problem by H. Rademacher [25]. It was verified by A. Brauer [26], using an induction argument based on the first of the two results in Exercise 3.3. The equivalent formula for $N(n,r,s)$ in terms of Ramanujan sums, i.e., Theorem 3.3 itself, was discovered by K. G. Ramanathan [44]. He also made the observations contained in Exercises 3.4 and 3.5.

Theorem 3.3 has been rediscovered and reproved several times. Once by C. A. Nicol and H. S. Vandiver [54], and again by D. Rearick [70] using the first result in Exercise 3.3, and by J. D. Dixon [60] using the second result in that exercise. E. Cohen [55b] observed that $N(\cdot,r,s)$ is an even function (mod r), and based a proof of Theorem 3.3 on this fact.

The Nagell function was studied by T. Nagell [23], and Proposition 3.4 was proved by E. Cohen [60j].

The generalization of Theorem 3.3 in Exercise 3.13 was obtained by E. Cohen [56a], and it is his argument that we have used to proof Theorem 3.3. The formula for $N_k(n,r,s)$ analogous to the Rademacher formula for $N(n,r,s)$ was obtained by L. Vietoris [67], and the latter obtain the result in Exercise 3.14.

The formulas for $P_k(n,r,s)$ and $Q_k(n,r,s)$ in the second and third examples of applications of Theorem 3.5 were found by E. Cohen [59e]. The formula for $P_2(n,r,s)$ in Exercise 3.12, as well as the result in Exercise 3.10, are also due to E. Cohen [59b].

The result in Exercise 3.8 was obtained by K. G. Ramanathan [44], its k-analogue by K. Nageswara Rao [67a] and the corresponding formula of the Rademacher type by L. Vietoris [68]. The generalization $D_{k,q}(n,r)$ of $c(n,r)$ defined in Exercise 3.9 was introduced by M. V. Subbarao and V. C. Harris [66], who obtained the formula for $P_{k,q}(n,r,s)$ given in that exercise.

E. Cohen [60e] counted the solutions of systems of linear congruences under various restrictions. The definition of $c(n_1,\ldots,n_t, r)$, and the evaluations of $N(m,n,r,s)$ and $\theta(m,n,r)$ at the end of the chapter and the results in Exercises 3.23-3.28 are in his paper.

The unified approach to evaluating numbers of restricted solutions of linear congruences that is embodied in Theorems 3.5 and 3.9 is due to P. J. McCarthy [75], [77]. The formula for $N_k(n,r,s)$ in the fourth example of an application of Theorem 3.5 can be found in the second paper.

The number $N'(n,r,s)$ of Theorem 3.7 was evaluated as it is in that theorem by E. Cohen [59a], and he obtained the formulas in Exercises 3.16 and 3.17. The proof of Theorem 3.7 using the Inclusion-Exclusion Principle was given by P. J. McCarthy [62] who proved, in fact, the k-analoge of Theorem 3.7: see also the paper of M. Suganamma [60]. Proposition 3.1, which is used in the proof of Theorem 3.7, is due to D. N. Lehmer [13].

E. Cohen [55a] proved (the k-analogue of) Theorem 3.8, and he gave two more proofs of the same result in another paper [56b]. In still another paper [59a], he gave one more proof using the fact that $S(\cdot,r,s)$ is an even function (mod r).

Chapter 4
Generalizations of Dirichlet Convolution

Let K be a complex-valued function on the set of all ordered pairs $\langle n,d\rangle$ where n is a positive integer and d is a divisor of n. If f and g are arithmetical functions, their <u>K-convolution</u>, $f *_K g$, is defined by

$$(f *_K g)(n) = \sum_{d|n} K(n,d)f(d)g(n/d) \quad \text{for all } n .$$

For example, if $K(n,d) = 1$ for all n and d, then $f *_K g = f * g$, the Dirichlet convolution of f and g.

We begin this chapter by investigating the binary operation $*_K$ on the set of arithmetical functions. We ask under what conditions the set of arithmetical functions is a commutative ring with respect to addition and K-convolution, with δ as unity.

Note that for every arithmetical function f, and for all n,

$$(f *_K \delta)(n) = K(n,n)f(n)$$

and

$$(\delta *_K f)(n) = K(n,1)f(n) .$$

If we take $f = \zeta$, then $\zeta *_K \delta = \zeta$ and $\delta *_K \zeta = \zeta$ imply that

(1) $$K(n,n) = K(n,1) = 1 \quad \text{for all } n .$$

On the other hand, it is certainly true that if (1) holds then $f *_K \delta = f$

and $\delta *_K f = f$ for every arithmetical function f.

Before investigating the associativity and communtativity of K-convolution, we show that the K-convolution of multiplicative functions is multiplicative if and only if

(2) $K(mn,de) = K(m,d)K(n,e)$ for all m,n,d and e such that $d|m$, $e|n$ and $(m,n) = 1$.

Assume $(m,n) = 1$, let $d|m$ and $e|n$, and define f and g by

$$f(N) = \begin{cases} 1 & \text{if } N|de \\ 0 & \text{otherwise} \end{cases} \qquad \text{for all } N$$

and

$$g(N) = \begin{cases} 1 & \text{if } N|mn/de \\ 0 & \text{otherwise} \end{cases} \qquad \text{for all } N .$$

Then f and g are multiplicative functions. If $f *_K g$ is multiplicative then

$$(f *_K g)(m)(f *_K g)(n) = K(m,d)K(n,e)$$

and

$$(f *_K g)(mn) = K(mn,de)$$

are equal. Therefore, (2) holds.

Conversely, if (2) holds and if f and g are multiplicative functions, then for all m and n with $(m,n) = 1$,

$$(f *_K g)(mn) = \sum_{d|mn} K(mn,d)f(d)g(mn/d)$$

$$= \sum_{\substack{d_1|m \\ d_2|n}} K(mn,d_1d_2)f(d_1)f(d_2)g(m/d_1)g(n/d_2)$$

$$= \left(\sum_{d_1|m} K(m,d_1)f(d_1)g(m/d_1)\right)\left(\sum_{d_2|n} K(n,d_2)f(d_2)g(n/d_2)\right)$$

$$= (f *_K g)(m)(f *_K g)(n) .$$

Thus, $f *_K g$ is multiplicative.

Now suppose that K-convolution is an associative binary operation. Let n be a positive integer and suppose $d|n$ and $e|d$. Define arithmetical functions f, g and h by

$$f(N) = \begin{cases} 1 & \text{if } N = e \\ 0 & \text{otherwise} \end{cases} \qquad \text{for all } N ,$$

$$g(N) = \begin{cases} 1 & \text{if } N = d/e \\ 0 & \text{otherwise} \end{cases} \qquad \text{for all } N$$

and

$$h(N) = \begin{cases} 1 & \text{if } N = n/d \\ 0 & \text{otherwise} \end{cases} \qquad \text{for all } N .$$

Then

$$((f *_K g) *_K h)(n) = K(n,d)K(d,e)$$

and

$$(f *_K (g *_K h))(n) = K(n,e)K(n/e,d/e) .$$

This proves the necessity of the condition in the statement that K-convolution is associative if and only if

(3) $K(n,d)K(d,e) = K(n,e)K(n/e,d/e)$ for all n, d and e such that $d|n$ and $e|d$.

On the other hand, if (3) holds then for all arithmetical functions f, g and h , and all n ,

$$((f *_K g) *_K h)(n) = \sum_{d|n} \sum_{e|d} K(d,e)K(n,d)f(e)g(d/e)h(n/d)$$

$$= \sum_{e|n} \sum_{\frac{d}{e}|\frac{n}{e}} K(n,e)K(n/e,d/e)f(e)g(d/e)h((n/e)/(d/e))$$

$$= (f *_K (g *_K h))(n) .$$

Thus, K-convolution is associative.

It can be shown by a similar argument that K-convolution is commutative if and only if

(4) $K(n,d) = K(n,n/d)$ for all n and d such that $d|n$:

this is Exercise 4.1.

It is clear that for all arithmetical functions f, g and h,

$$f *_K (g + h) = f *_K g + f *_K h$$

and

$$(g + h) *_K f = g *_K f + h *_K f .$$

Therefore, the set of arithmetical functions is a commutative ring with respect to addition and K-convolution, with unity δ, in which the product (i.e., K-convolution) of multiplicative functions is multiplicative, if and only if (1)-(4) hold.

Suppose that (1)-(4) hold for K, and consider the ring of arithmetical functions (with respect to addition and K-convolution). If an arithmetical function f has an inverse f^{-1} in this ring then

$$1 = (f *_K f^{-1})(1) = K(1,1)f(1)f^{-1}(1) = f(1)f^{-1}(1) .$$

Thus,

$$f(1) \neq 0 \quad \text{and} \quad f^{-1}(1) = \frac{1}{f(1)} .$$

Conversely, suppose $f(1) \neq 0$ and define f^{-1} by

$$f^{-1}(1) = \frac{1}{f(1)}$$

and

$$f^{-1}(n) = -\sum_{\substack{d|n \\ d>1}} K(n,d)f(d)f^{-1}(n/d) \quad \text{for all} \quad n > 1 .$$

Then $f *_K f^{-1} = \delta$ and, since K-convolution is associative, $f^{-1} *_K f = \delta$.

We leave it as Exercise 4.4 to show that if f is a multiplicative function and $f(1) \neq 0$, then f^{-1} is multiplicative.

Now consider the following example of a K-convolution. For all n and d such that $d|n$ let

$$K(n,d) = \begin{cases} 1 & \text{if } (d,n/d) = 1 \\ 0 & \text{otherwise} . \end{cases}$$

A divisor d of n such that $(d,n/d) = 1$ is called a unitary divisor. To denote the fact that d is a unitary divisor of n we will write $d \| n$. If f and g are arithmetical functions then for all n,

$$(f *_K g)(n) = \sum_{d \| n} f(d)g(n/d) .$$

Certainly (1), (2) and (4) hold for K. To verify that (3) also holds we must show that for all n, d and e such that $d|n$ and $e|d$, the statements "$(d,n/d) = 1$ and $(e,d/e) = 1$" and "$(e,n/e) = 1$ and $(d/e,(n/e)/(d/e)) = 1$" are equivalent. However, because (2) holds it is only necessary to verify this equivalence when $n = p^\alpha$, $d = p^\beta$ and $e = p^\gamma$, where p is a prime and $0 \leq \gamma \leq \beta \leq \alpha$, in which case the two statements become, respectively,

$$\text{"}(\beta = \alpha \text{ or } \beta = 0) \text{ and } (\gamma = \beta \text{ or } \gamma = 0)\text{"}$$

and

$$"(\gamma = \alpha \quad \text{or} \quad \gamma = 0) \quad \text{and} \quad (\beta = \alpha \quad \text{or} \quad \beta = \gamma)" .$$

A quick check shows that these statements are equivalent.

Continuing with the same K, let $\mu_K = \zeta^{-1}$. Then

$$\sum_{d \| n} \mu_K(d) = \begin{cases} 1 & \text{if } n = 1 \\ 0 & \text{otherwise} \end{cases}$$

If p is a prime and $\alpha \geq 1$ then $\mu_K(1) + \mu_K(p^\alpha) = 0$, i.e., $\mu_K(p^\alpha) = -1$. Therefore, the K-convolution for this particular K, which is called the unitary convolution, is regular in the sense of the following definition.

A K-convolution is called a regular arithmetical convolution if the following conditions are satisfied:

(a) (1)-(4) hold for K.

(b) $K(n,d) = 0$ or 1 for all n and d such that $d|n$.

(c) If $\mu_K = \zeta^{-1}$ then $\mu_K(p^\alpha) = 0$ or -1 for all primes p and all $\alpha \geq 0$.

The Dirichlet and unitary convolutions are regular.

Suppose that K-convolution is a regular arithmetical convolution. For each n let

$$A(n) = \{d : d|n \text{ and } K(n,d) = 1\} .$$

Then we shall refer to the "regular arithmetical convolution A," and write $*_A$ and μ_A in place of $*_K$ and μ_K. If f and g are arithmetical functions then

$$(f *_A g)(n) = \sum_{d \in A(n)} f(d)g(n/d) \qquad \text{for all } n .$$

On the other hand, suppose that for each positive integer n we choose a nonempty subset $A(n)$ of the set of divisors of n. For all n and d such that $d|n$ let

$$K(n,d) = \begin{cases} 1 & \text{if } d \in A(n) \\ 0 & \text{if } d \notin A(n) . \end{cases}$$

The resulting K-convolution may or may not be a regular arithmetical convolution. The conditions (1)-(4) are equivalent to

(1') $1, n \in A(n)$ for all n,

(2') if $(m,n) = 1$ then

$$A(mn) = \{de : d \in A(m) \text{ and } e \in A(n)\} ,$$

(3') the statements "$d \in A(n)$ and $e \in A(d)$" and "$e \in A(n)$ and $d/e \in A(n/e)$" are equivalent

and

(4') if $d \in A(n)$ then $n/d \in A(n)$,

respectively.

Theorem 4.1. If a K-convolution satisfies (a) and (b) then it satisfies (c) if and only if for each prime p and each $\alpha \geq 1$ there is a divisor t of α such that

$$A(p^\alpha) = \{1, p^t, p^{2t}, \ldots, p^\alpha\}$$

and $A(p^{ht}) = \{1, p^t, p^{2t}, \ldots, p^{ht}\}$ for $h = 1, \ldots, \alpha/t$.

Proof. Assume that the K-convolution satisfies (a), (b) and (c). Let

$$A(p^\alpha) = \{1, p^{\alpha_1}, p^{\alpha_2}, \ldots, p^{\alpha_k}\} , \alpha_k = \alpha,$$

where $0 < \alpha_1 < \alpha_2 < \ldots < \alpha_k$. Suppose $p^j \in A(p^{\alpha_1})$ for some j with $0 < j < \alpha_1$. Then, by (3'), $p^j \in A(p^\alpha)$ which is not true. Hence, $A(p^{\alpha_1}) = \{1, p^{\alpha_1}\}$. Since $\mu_A = \zeta^{-1}$, the inverse of ζ with respect to the binary operation $*_A$, we have $\mu_A(1) = 1$ and

$$0 = 1 + \mu_A(p^{\alpha_1}) , \text{ i.e., } \mu_A(p^{\alpha_1}) = -1 .$$

Then

$$0 = 1 + \mu_A(p^{\alpha_1}) + \mu_A(p^{\alpha_2}) + \ldots + \mu_A(p^{\alpha_k})$$

$$= \mu_A(p^{\alpha_2}) + \ldots + \mu_A(p^{\alpha_k})$$

which, by (c), implies that $\mu_A(p^{\alpha_i}) = 0$ for $i = 2, \ldots, k$. Let $t = \alpha_1$.

We shall continue by induction to show that for $h = 1, \ldots, k$,

$$\alpha_h = ht \quad \text{and} \quad A(p^{ht}) = \{1, p, p^2, \ldots, p^{ht}\} :$$

it will follow that $\alpha = kt$ and

$$A(p^\alpha) = \{1, p^t, p^{2t}, \ldots, p^\alpha\}.$$

Assume that $h > 1$ and that for $j = 1, \ldots, h-1$,

$$\alpha_j = jt \quad \text{and} \quad A(p^{jt}) = \{1, p^t, p^{2t}, \ldots, p^{jt}\} .$$

By (3'), $A(p^{\alpha_h})$ is subset of $\{1, p^t, \ldots, p^{(h-1)t}, p^{\alpha_h}\}$. Since $\mu_A(p^{\alpha_i}) = 0$ for $i = 2, \ldots, h$, $A(p^{\alpha_h})$ must contain p^t. By (4'), $p^{\alpha_h - t} \in A(p^{\alpha_h})$ and since $\alpha_h > \alpha_h - t > (h-2)t$ we have $\alpha_h - t = (h-1)t$, i.e., $\alpha_h = ht$ and $p^{(h-1)t} \in A(p^{\alpha_h})$. Now let $2 \le i \le h - 2$. If $n = p^{ht}$, $d = p^{it}$ and $e = p^t$, then

$$d/e = p^{(i-1)t} \in A(p^{(h-1)t}) = A(n/e) \quad \text{and} \quad e = p^t \in A(p^{ht}) = A(n) .$$

hence, by (3'),

$$p^{it} = d \in A(n) = A(p^{ht}) .$$

Therefore,

$$A(p^{\alpha_h}) = A(p^{ht}) = \{1, p, p^2, \ldots, p^{ht}\} .$$

Conversely, suppose (a) and (b) hold. Let p be a prime and $\alpha \ge 1$ and suppose that

$$A(p^\alpha) = \{1, p^t, p^{2t}, \ldots, p^\alpha\}$$

where $t|\alpha$, and $A(p^{ht}) = \{1, p^t, p^{2t}, \ldots, p^{ht}\}$ for $h = 1, \ldots, \alpha/t$. Since $A(p^t) = \{1, p^t\}$ we have $0 = 1 + \mu_A(p^t)$, i.e., $\mu_A(p^t) = -1$. Suppose $\alpha > t$. Since $A(p^{2t}) = \{1, p^t, p^{2t}\}$ we have $0 = 1 + \mu_A(p^t) + \mu_A(p^{2t}) = \mu_A(p^{2t})$. In the same manner, we can show that $\mu_A(p^{3t}) = 0$ and so on, until we conclude that $\mu_A(p^\alpha) = 0$. Thus (c) holds. □

Consider a regular arithmetical convolution A. If p is a prime and $\alpha \ge 1$ then

$$A(p^\alpha) = \{1, p^t, p^{2t}, \ldots, p^{kt}\} , \ \alpha = kt .$$

The divisor t of α is called the <u>type of</u> p^α <u>with respect to</u> A : it will be denoted by $t_A(p^\alpha)$.

<u>Corollary 4.2</u>. Let p be a prime and $\alpha, \beta \geq 1$. If $A(p^\alpha) \cap A(p^\beta) \neq \{1\}$ then $t_A(p^\alpha) = t_A(p^\beta)$ (say = t) and (assuming $\alpha \leq \beta$) $A(p^\alpha)$ consists of the $(\alpha/t) + 1$ smallest integers in $A(p^\beta)$.

Proof. If $p^\gamma \in A(p^\alpha) \cap A(p^\beta)$ and $p^\gamma > 1$ then $t_A(p^\alpha) = t_A(p^\gamma) = t_A(p^\beta)$. Then for some h and k , $\alpha = ht$ and $\beta = kt$, and if $\alpha \leq \beta$,

$$A(p^\alpha) = \{1, p^t, p^{2t}, \ldots, p^{ht}\} \text{ and}$$

$$A(p^\beta) = A(p^\alpha) \cup \{p^{(h+1)t}, \ldots, p^{kt}\} . \square$$

It is clear that a regular arithmetical convolution A is completely determined by the sets $A(p^\alpha)$ for all primes p and all $\alpha \geq 1$. These sets are perfectly arbitrary within the constraints set down by Theorem 4.1.

Thus, if $A_1, \ldots, A_n$ are regular arithmetical convolutions and if $S_1 \cup \ldots \cup S_h$ is any partition of the set of primes into h sets, then there is a unique regular arithmetical convolution A such that for $i = 1, \ldots, h,$

$$A(p^\alpha) = A_i(p^\alpha) \quad \text{for all } p \in S_i \text{ and all } \alpha \geq 1 .$$

The Dirichlet convolution will be denoted by D . For all p^α,

$$D(p^\alpha) = \{1, p, p^2, \ldots, p^\alpha\}$$

and $t_D(p^\alpha) = 1$.

The unitary convolution will be denoted by U . For all p^α ,

$$U(p^\alpha) = \{1, p^\alpha\}$$

and $t_U(p^\alpha) = \alpha$.

If A is an arbitrary arithmetical convolution then for all p^α,

$$U(p^\alpha) \subseteq A(p^\alpha) \subseteq D(p^\alpha) .$$

Thus, in some sense the Dirichlet and unitary convolutions are the "extreme" convolutions. This is made more precise in Exercise 4.7.

Let A be a regular arithmetical convolution. A positive integer n for which $A(n) = \{1, n\}$ is called primitive (with respect to A). By (2'), if n is primitive then $n = p^\alpha$ for some prime p and $\alpha \geq 1$. Furthermore, p^α is primitive if and only if $t_A(p^\alpha) = \alpha$. It is shown in the proof of Theorem 4.1 that if p is a prime and $\alpha \geq 1$, then

$$\mu_A(p^\alpha) = \begin{cases} -1 & \text{if } p^\alpha \text{ is primitive} \\ 0 & \text{otherwise} . \end{cases}$$

A regular arithmetical convolution is not determined uniquely by the corresponding primitive integers. For example, let A and B be the regular arithmetical convolutions for which

$A(p^\alpha) = B(p^\alpha) = \{1, p^\alpha\}$ for every odd prime p and all $\alpha \geq 1$,

$A(2^\alpha) = B(2^\alpha) = \{1, 2^\alpha\}$ for all $\alpha \neq 6, 8, 9. 12,$

$A(2^6) = B(2^6) = \{1, 2^3, 2^6\}$,

$A(2^8) = B(2^8) = \{1, 2^4, 2^8\}$,

$A(2^9) = B(2^9) = \{1, 2^3, 2^6, 2^9\}$,

$A(2^{12}) = \{1, 2^3, 2^6, 2^9, 2^{12}\}$,

$B(2^{12}) = \{1, 2^4, 2^8, 2^{12}\}$.

Then $A \neq B$ but the primitive integers with respect to A are the same as the primitive integers with respect to B .

Let A be a regular arithmetical convolution. From the definition of μ_A as the inverse of ζ there follows an analogue of the Möbius inversion theorem: if f and g are arithmetical functions then

$$f(n) = \sum_{d \in A(n)} g(d) \quad \text{for all } n$$

if and only if

$$g(n) = \sum_{d \in A(n)} f(d)\mu_A(n/d) \quad \text{for all } n .$$

If a and b are integers then $(a,b)_A$ will denote the largest divisor d of a such that $d \in A(b)$. Thus,

$(a,b)_D = (a,b)$, the greatest common divisor of a and b ,

and

$(a,b)_U =$ the largest integer d such that $d|a$ and $d \| b$.

An analogue of the Euler function, which will be denoted by ϕ_A , is defined by

$\phi_A(n)$ = the number of integers x such that $1 \le x \le n$ and $(x,n)_A = 1$.

To evaluate $\phi_A(n)$ we will use the following result, which extends Lemma 1.4 to an arbitrary regular arithmetical convolution.

Lemma 4.3. For each $d \in A(n)$ let

$$S_d = \{xn/d : 1 \le x \le d \text{ and } (x,d)_A = 1\}.$$

If $d \ne e$ then $S_d \cap S_e$ is empty, and

$$\bigcup_{d \in A(n)} S_d = \{1, \ldots, n\} .$$

Proof. Let $d, e \in A(n)$, $1 \le x \le d$, $(x,d)_A = 1$, $1 \le y \le e$ and $(y,e)_A = 1$, and suppose that $xn/d = yn/e$, i.e., $xe = yd$. We shall show that $x = y$, and therefore $d = e$. It is enough to show that if p is a prime divisor of x and if p^u is the highest power of p that divides x , then $p^u | y$. If $p \nmid d$ then $p^u | y$.

On the other hand, suppose that $p|d$. Then $p|n$: let p^α , p^β and p^γ be the highest powers of p that divide n, d and e , respectively. Note that we may have $\gamma = 0$. Let $t = t_A(p^\alpha)$. Now

$$d \in A(n) \quad \text{and} \quad p^\beta \in A(d)$$

imply $p^\beta \in A(n)$ which, in turn, implies that $p^\beta \in A(p^\alpha)$ (we have used (3') and Exercise 4.5). Hence $\beta = it$ for some $i > 0$, and $t_A(p^\beta) = t$ by Theorem 4.1. Likewise $\gamma = jt$ for some $j \ge 0$, and if $j > 0$ then $t_A(p^\gamma) = t$. Since $(x,d)_A = 1$, $u < t$. Also, if $j > 0$ and if p^v

is the highest power of p that divides y, then $v < t$ since $(y,e)_A = 1$. Now, $p^u p^{jt} = p^v p^{it}$: hence $j > 0$ for otherwise $t \leq u$. Since $u - v = (i-j)t$ it follows that $i = j$, and $u = v$.

It remains to show that if $1 \leq m \leq n$ then $m \in S_d$ for some $d \in A(n)$. In fact, let d be such that $(m,n)_A = n/d$: then $d \in A(n)$ by (4'). Let $m = xn/d$ and let $e = (x,d)_A$. Then $d \in A(n)$ and $e \in A(d)$ imply $e \in A(n)$ and $d/e \in A(n/e)$ by (3'). Since $n/e \in A(n)$, we conclude that $d|e \in A(n)$. However, $m = (x/e)(en/d)$, i.e., $e(n/d)|m$. Therefore, by the definition of d, $e = 1$. □

It follows immediately that

$$n = \sum_{d \in A(n)} \phi_A(d) \quad \text{for all } n ,$$

and therefore

$$\phi_A(n) = \sum_{d \in A(n)} d\, \mu_A(n/d) \quad \text{for all } n .$$

Since μ_A is multiplicative, ϕ_A is a multiplicative function. If p is a prime and $\alpha \geq 1$, then

$$\phi_A(p^\alpha) = p^\alpha - p^{\alpha-t} , \quad t = t_A(p^\alpha) .$$

For example, if p is a prime and $\alpha \geq 1$ then

$$\phi_U(p^\alpha) = p^\alpha - 1 .$$

Thus, for all n,

$$\phi_U(n) = \prod_{p^\alpha \| n} (p^\alpha - 1) .$$

The analogue of the Ramanujan sum $c(n,r)$ for a regular arithmetical convolution A is defined as follows: let r be a positive integer and let

$$c_A(n,r) = \sum_{(x,r)_A = 1} e(nx,r) \quad \text{for all } n .$$

By Lemma 4.3,

$$\sum_{h=1}^{r} e(nh,r) = \sum_{d \in A(r)} \sum_{(x,d)_A = 1} e(nxr/d,r)$$

$$= \sum_{d \in A(r)} \sum_{(x,d)_A = 1} e(nx,d)$$

$$= \sum_{d \in A(r)} c_A(n,d) ,$$

and for fixed n this holds for all r. Thus, by the inversion theorem

$$c_A(n,r) = \sum_{d|n, d \in A(r)} d\, \mu_A(r/d) .$$

It follows immediately that

$$c_A(n,r) = \phi_A(r) \quad \text{if} \quad r|n$$

and

$$c_A(n,r) = \mu_A(r) \quad \text{if} \quad (n,r)_A = 1 \ .$$

If, for positive integers a and b, we let

$$\varepsilon_A(a,b) = \begin{cases} 1 & \text{if} \quad a \in A(ab) \\ 0 & \text{if} \quad a \notin A(ab) \end{cases}$$

and

$$g(a,b) = a\,\mu_A(b)\varepsilon_A(a,b) \ ,$$

then for every divisor d of r,

$$g(d,r/d) = \begin{cases} d\,\mu_A(r/d) & \text{if} \quad d \in A(r) \\ 0 & \text{if} \quad d \notin A(r) \ . \end{cases}$$

Thus,

$$c_A(n,r) = \sum_{d|(n,r)} g(d,r/d) \ ,$$

and it follows by Proposition 2.10 that $c_A(\cdot,r)$ is an even function (mod r).

Let

$$c_A(n,r) = \sum_{d|r} \alpha(d)c(n,d) \quad \text{for all} \quad n \ .$$

Again from Proposition 2.10,

$$\alpha(d) = \frac{1}{r} \sum_{e|\frac{r}{d}} g(r/e,e)e$$

$$= \sum_{e|\frac{r}{d}} \mu_A(e)\varepsilon_A(r/e,e) .$$

Let $r = p_1^{\alpha_1} \cdots p_h^{\alpha_h}$ where each $\alpha_i > 0$, and $d = p_1^{\beta_1} \cdots p_h^{\beta_h}$ where for $i = 1, \ldots, h$, $0 \leq \beta_i \leq \alpha_i$. Then, because of (2') and the fact that μ_A is a multiplicative function,

$$\alpha(d) = \prod_{i=1}^{h} \sum_{j=0}^{\alpha_i-\beta_i} \mu_A(p_i^{\,j})\varepsilon_A(p_i^{\alpha_i-j}, p_i^{\,j}) .$$

Now $\varepsilon_A(p_i^{\alpha_i-j}, p_i^{\,j}) = 1$ if and only if $p_i^{\alpha_i-j} \in A(p_i^{\alpha_i})$, and by Theorem 4.1 this is true if and only if $t|\alpha_i-j$, where $t = t_A(p_i^{\alpha_i})$, i.e., if and only if $t|j$. Thus,

$$\sum_{j=0}^{\alpha_i-\beta_i} \mu_A(p_i^{\,j})\varepsilon_A(p_i^{\alpha_i-j}, p_i^{\,j}) = \begin{cases} 1 & \text{if } \alpha_i-\beta_i<t, \text{ i.e., } \beta_i \geq \alpha_i-t+1 \\ 1+\mu_A(p_i^{\,t})=0 & \text{if } \alpha_i-\beta_i \geq t, \text{ i.e., } \beta_i<\alpha_i-t+1. \end{cases}$$

To express $\alpha(d)$ efficiently we define the <u>A-core function</u> to be the multiplicative function γ_A such that for each prime p and all $\alpha \geq 1$,

$$\gamma_A(p^\alpha) = p^{\alpha-t+1}, \text{ where } t = t_A(p^\alpha) .$$

Note that $\gamma_D(p^\alpha) = p^\alpha$, so that $\gamma_D(n) = n$ for all n , and $\gamma_U(p^\alpha) = p$, so that $\gamma_U(n) = \gamma(n)$ for all n, where γ is the core function defined in Exercise 1.14.

For each divisor d of r,

$$\alpha(d) = \begin{cases} 1 & \text{if } \gamma_A(r) \mid d \\ 0 & \text{if } \gamma_A(r) \nmid d \ . \end{cases}$$

Therefore,

$$c_A(n,r) = \sum_{d \mid r,\ \gamma_A(r) \mid d} c(n,d) \quad \text{for all } n \ .$$

In particular,

$$\phi_A(r) = \sum_{d \mid r,\ \gamma_A(r) \mid d} \phi(d)$$

and

$$\mu_A(r) = \sum_{d \mid r,\ \gamma_A(r) \mid d} \mu(d) \ .$$

We can apply Theorem 3.5 to obtain the number $N_A(n,r,s)$ of solutions $\langle x_1, \ldots, x_s \rangle$ of the congruence (*) of Chapter 3 such that each $(x_i,r)_A = 1$. If each $T_i(r) = \{x : 1 \leq x \leq r \text{ and } (x,r)_A = 1\}$ then each $g_i(n,r) = c_A(n,r)$.

<u>Theorem 4.4.</u> For all n,

$$N_A(n,r,s) = \frac{1}{r} \sum_{d \mid r} c_A(r/d,r)^s c(n,d) \ .$$

Another formula for $N_A(n,r,s)$ is given in Exercise 4.35

Exercises for Chapter 4

4.1. K-convolution is commutative if and only if (4) holds.

4.2. Let g be an arithmetical function, and for all n and d such that $d|n$ let

$$K(n,d) = \frac{g(n)}{g(d)g(n/d)} .$$

Then (1), (3) and (4) hold, and so does (2) if g is multiplicative.

4.3. For a prime p and integers α and β such that $0 \leq \beta \leq \alpha$, let

$$K(p^\alpha,p^\beta) = \binom{\alpha}{\beta} ,$$

and then define $K(n,d)$ for all n and d such that $d|n$ by requiring that (2) holds. Then (1)-(4) hold, and the K-convolution of two completely multiplicative functions is completely multiplicative.

4.4. Suppose (1)-(4) hold for K. If f is a multiplicative function and $f(1) \neq 0$, then f^{-1}, the inverse of f with respect to K-convolution, is multiplicative.

4.5. Let A be a regular arithmetical convolution. If p is a prime and p^α is the highest power of p that divides a positive integer n, then $p^\alpha \in A(n)$. Furthermore, if $p^\beta \in A(n)$ then $p^\beta \in A(p^\alpha)$.

4.6. Let $S = \{p : p$ is a prime and $p \equiv 1$ or $2 \pmod 4\}$ and $T = \{p : p$ is a prime and $p \equiv 3 \pmod 4\}$. Let A be the regular arithmetical convolution such that for all $\alpha \geq 1$,

$$A(p^\alpha) = \begin{cases} U(p^\alpha) & \text{if } p \in S \\ D(p^\alpha) & \text{if } p \in T . \end{cases}$$

Then $\mu_A(n) \neq 0$ if and only if for every prime $p \equiv 3 \pmod 4$, if $p|n$ then $p^2 \nmid n$.

4.7. Let A and B be regular arithmetical convolutions. The following statements are equivalent:

(a) $(m,n)_B = ((m,n)_A,n)_B$ for all nonnegative integers m and positive integers n .

(b) $t_A(p^\alpha) | t_B(p^\alpha)$ for all primes p and all $\alpha \geq 1$.

(c) $B(p^\alpha) \subseteq A(p^\alpha)$ for all primes p and all $\alpha \geq 1$.

If these statements hold we write $B \leq A$. Then $\leq$ is a partial ordering on the set of all regular arithmetical convolutions. If C is an arbitrary regular arithmetical convolution then $U \leq C \leq D$.

4.8. Let A be a regular arithmetical convolution and let p be a prime and $\alpha \geq 1$. Let $t = t_A(p^\alpha)$. For all n ,

$$c_A(n,p^\alpha) = \begin{cases} p^\alpha - p^{\alpha-t} & \text{if } p^\alpha | n \\ -p^{\alpha-t} & \text{if } p^{\alpha-t} | n,\ p^\alpha \nmid n , \\ 0 & \text{if } p^{\alpha-t} \nmid n . \end{cases}$$

Thus, $c_A(n,p^\alpha) = c_t(n,p^{\alpha/t})$. (The extended Ramanujan sum $c_k(n,r)$ is defined in Exercise 2.51.)

4.9. Let A be a regular arithmetical convolution and r a positive integer. For all n,

$$c_A(n,r) = \frac{\phi_A(r)\mu_A(m)}{\phi_A(m)}, \quad m = \frac{r}{(n,r)_A}.$$

(Hint: use Exercises 2.55 and 4.8.)

4.10. Let A be a regular arithmetical convolution. If r is a positive integer and $d, e \in A(r)$ then for all n,

$$\sum_{n\equiv a+b \pmod r} c_A(a,d)c_A(b,e) = \begin{cases} r\, c_A(n,d) & \text{if } d = e \\ 0 & \text{if } d \neq e \end{cases}$$

4.11. Continuing Exercise 4.10,

$$\sum_{D\in A(r)} \frac{c_A(d,D)c_A(e,D)}{\phi_A(D)} = \begin{cases} \dfrac{r}{\phi_A(r/d)} & \text{if } d = e \\ 0 & \text{if } d \neq e \end{cases}$$

4.12. Continuing Exercise 4.11,

$$\sum_{D\in A(r)} c_A(r/D,d)c_A(r/e,D) = \begin{cases} r & \text{if } d = e \\ 0 & \text{if } d \neq e \end{cases}$$

4.13. Let A be a regular arithmetical convolution and r a positive integer. An arithmetical function f is called an <u>A-even function (mod r)</u> if $f(n) = f((n,r)_A)$ for all n. If f is such a function then

$$f(n) = \sum_{d \in A(r)} \alpha_A(d) c_A(n,d) \quad \text{for all } n ,$$

where the coefficients $\alpha_A(d)$ are uniquely determined by f and are given by

$$\alpha_A(d) = \frac{1}{r} \sum_{e \in A(r)} f(r/e,r) c_A(r/d,e) .$$

4.14. Continuing Exercise 4.13, an arithmetical function f is A-even (mod r) if and only if there is a complex-valued function g of two positive integer variables such that for all n,

$$f(n,r) = \sum_{d|n, d \in A(r)} g(d,r/d) .$$

In this case, for all $d \in A(r)$,

$$\alpha_A(d) = \frac{1}{r} \sum_{e \in A(r/d)} g(r/e,e)e .$$

4.15. Let A and B be regular arithmetical convolutions. Then $B \leq A$ if and only if every B-even arithmetical function (mod r) is an A-even function (mod r).

4.16. If A and B are regular arithmetical convolutions and if $B \leq A$ then for all n and r,

$$c_B(n,r) = \sum_{d \in A(r),\ \gamma_B(r)|d} c_A(n,d) .$$

In particular, for all A,

$$c_U(n,r) = \sum_{d \in A(r), \gamma(r)=\gamma(d)} c_A(n,d) .$$

4.17. Let A be a regular arithmetical convolution. If f and g are arithmetical functions then

$$f(n) = \sum_{d \in A(n), \gamma(d)=\gamma(n)} g(d) \quad \text{for all } n$$

if and only if

$$g(n) = \sum_{d \in A(n), \gamma(d)=\gamma(n)} f(d)\mu_A(n/d) \quad \text{for all } n$$

Thus, for all n and r,

$$c_A(n,r) = \sum_{d \in A(r), \gamma(d)=\gamma(r)} c_U(n,d)\mu_A(r/d) .$$

4.18. Let A be a regular arithmetical convolution. An arithmetical function f is <u>A-multiplicative</u> if for all n,

$$f(d)f(n/d) = f(n) \quad \text{for all } d \in A(n) .$$

(Thus, f is D-multiplicative if and only if it is completely multiplicative.) If f is A-multiplicative then it is a multiplicative function. On the other hand, if f is a multiplicative function then f is A-multiplicative if and only if

$$f(p^\alpha) = f(p^t)^{\alpha/t} \quad , \; t = t_A(p^\alpha) \; ,$$

for every prime p and all $\alpha \geq 1$.

4.19. Continuing Exercise 4.18, an arithmetical function f is A-multiplicative if and only if $f(g *_A h) = fg *_A fh$ for all arithmetical functions g and h .

4.20. Continuing Exercise 4.19, a multiplicative function f is A-multiplicative if and only if $f^{-1} = f\mu_A$. (f^{-1} is the inverse of f with respect to $*_A$.)

4.21. Continuing Exercise 4.20, a multiplicative function f is A-multiplicative if and only if $f^{-1}(p^\alpha) = 0$ for every prime p and all $\alpha \geq 1$ such that p^α is not primitive.

4.22. Let A be a regular arithmetical convolution. If g_1, g_2, h_1 and h_2 are A-multiplicative functions then

$$g_1g_2 *_A g_1h_2 *_A h_1g_2 *_A h_1h_2 = (g_1 *_A h_1)(g_2 *_A h_2) *_A u,$$

where u is the multiplicative function such that for every prime p and all $\alpha \geq 1$,

$$u(p^\alpha) = \begin{cases} g_1g_2h_1h_2(p^{\alpha/2}) & \text{if } \alpha/t \text{ is even} \\ 0 & \text{if } \alpha/t \text{ is odd ,} \end{cases}$$

$t = t_A(p^\alpha)$. (See Exercise 1.63)

4.23. Let A be a regular arithmetical convolution. For each positive integer n, let $\tau_A(n) = \#A(n)$. Let k be a positive integer. If f is a multiplicative function then there exist A-multiplicative functions $g_1,\ldots,g_k$ such that $f = g_1 *_A \cdots *_A g_k$ if and only if $f^{-1}(p^\alpha) = 0$ for every prime p and all $\alpha \geq 1$ such that $\tau_A(p^\alpha) \geq k + 2$.

4.24. Continuing Exercise 4.23, suppose that $f = g_1 *_A \cdots *_A g_k$, where each function g_i is A-multiplicative. If $\tau_A(p^\alpha) = k + 1$ then $f^{-1}(p^\alpha) = (-1)^k g_1(p^t)\ldots g_k(p^t)$, where $t = t_A(p^\alpha)$.

4.25. Let A be a regular arithmetical convolution. An arithmetical function f is called an <u>A-specially multiplicative function</u> if $f = g_1 *_A g_2$, where g_1 and g_2 are A-multiplicative functions. By Exercise 4.23, a multiplicative function f is A-specially multiplicative if and only if $f^{-1}(p^\alpha) = 0$ for every prime p and all $\alpha \geq 1$ such that $\tau_A(p^\alpha) \geq 4$. A multiplicative function f is A-specially multiplicative if and only if for every prime p and all $\alpha \geq 1$ such that $\tau_A(p^\alpha) \geq 3$, and every k with $1 \leq k \leq \tau_A(p^\alpha) - 2$,

$$f(p^{(k+1)t}) = f(p^t)f(p^{kt}) + f(p^{(k-1)t})[f(p^{2t}) - f(p^t)^2].$$

4.26. Let K be a function of the kind considered at the beginning of the chapter, and assume that it satisfies conditions (1)-(4). Let V be the function defined in Exercise 1.67. If f is a multiplicative function then for all m and n,

$$f(mn) = \sum_{d|m} \sum_{e|n} f(m/d)f(n/e)f^{-1}(de)K(mn,mn/de)K(mn/de,m/d)V(d,e),$$

where f^{-1} is the inverse of f with respect to K-convolution.

4.27. Let A be a regular arithmetical convolution. If f is an A-specially multiplicative function, $f = g_1 *_A g_2$, where g_1 and g_2 are A-multiplicative functions, and if $m,n \in A(r)$ for some r, then

$$f(mn) = \sum_{d \in A((m,n))} f(m/d)f(n/d)g_1(d)g_2(d)\mu_A(d).$$

(Hint: use Exercises 4.26 and 4.24.)

4.28. Continuing Exercise 4.27, a multiplicative function f is A-specially multiplicative if and only if there is a multiplicative function F such that if $m,n \in A(r)$ for some r, then

$$f(mn) = \sum_{d \in A((m,n))} f(m/d)f(n/d)F(d).$$

This can be proved directly, without using Exercise 4.27.

4.29. Continuing Exercise 4.28, a multiplicative function f is A-specially multiplicative if and only if there is an A-multiplicative function B such that if $m,n \in A(r)$ for some r, then

$$f(m)f(n) = \sum_{d \in A((m,n))} f(mn/d^2)B(d).$$

4.30. Let A be a regular arithmetical convolution. Let h and g be multiplicative functions, and let

$$f(n,r) = \sum_{\substack{d|n \\ d\in A(r)}} h(d)g(r/d)\mu_A(r/d)$$

for all integers n and positive integers r. Let $F(r) = f(0,r)$ for all r, and assume that $F(r) \neq 0$ for all r. If h is A-multiplicative then a formula for $f(n,r)$ analogous to the one in Theorem 2.3 holds. (Hint: see Exercise 4.9.). If $g(p^\alpha) \neq 0$ for every primitive prime power p^α, then the converse statement is true.

4.31. Continuing Exercise 4.30, if h is A-multiplicative then an identity analogous to the one in Theorem 2.5 holds for $f(n,r)$ and F. If $h(p^\alpha) \neq 0 \neq g(p^\alpha)$ for every primitive prime power p^α, then the converse statement is true.

4.32. Continuing Exercise 4.31, if h is A-multiplicative then an identity analogous to the one in Exercise 2.66 holds for F. If $g(p^\alpha) \neq 0$ for every primitive prime power p^α, then the converse statement is true.

4.33. Let A be a regular arithmetical convolution. If B is the $n \times n$ matrix $[c_A(i,j)]$, $1 \leq i,j \leq n$, then $\det B = n!$.

4.34. Continuing Exercise 4.33, if C is the $n \times n$ matrix $[N_A(i,j,s)]$, $1 \leq i,j \leq n$, then $\det C = (\mu_A(1)\ldots\mu_A(n))^s$.

4.35. Let A be a regular arithmetical convolution, and let $N_A(n,r,s)$ be the number of solutions $\langle x_1,\ldots,x_s\rangle$ of the congruence (*) of Chapter 3 such that $(x_i,r)_A = 1$ for $i = 1,\ldots,s$.

For all n,

$$N_A(n,r,s) = \frac{1}{r} \sum_{d \in A(r)} c_A(r/d,r)^s \, c_A(n,d) \, .$$

This can be proved in the way that Theorem 3.3 was proved, and it can also be proved directly from Theorem 3.3 by using Exercise 4.8.

4.36. Continuing Exercise 4.35,

$$N_A(n,r,s) = r^{s-1} \prod_{\substack{p|r \\ p^t|n}} \frac{(p^t-1)((p^t-1)^{s-1}-(-1)^{s-1})}{p^{ts}} \prod_{\substack{p|r \\ p^t\nmid n}} \frac{(p^t-1)^s-(-1)^s}{p^{ts}} \, ,$$

where for each prime p that divides r, $t = \tau_A(p^\alpha)$, where $p^\alpha \| r$.

4.37. Continuing Exercise 4.36, $N_A(n,r,s) = 0$ if and only if either $s = 1$ and $(n,r)_A \neq 1$ or $s > 1$, $2 \in A(r)$ and $n \not\equiv s \pmod 2$. In particular, if the generalized Nagell function $\theta_A(\cdot,r)$ is defined by $\theta_A(n,r) = N_A(n,r,2)$, then $\theta_A(n,r) = 0$ if and only if n is odd and $2 \in A(r)$.

4.38. Let A be a regular arithmetical convolution, and let $N'_A(n,r,s)$ be the number of solutions $\langle x_1, \ldots, x_s \rangle$ of (*) such that $((x_1, \ldots, x_s),r)_A = 1$. For all n,

$$N'_A(n,r,s) = \sum_{d \in A((n,r)_A)} (r/d)^{s-1} \mu_A(d) \, .$$

(Hint: use the Inclusion-Exclusion Principle. Can this be proved using Exercise 4.8, as well?)

4.39. Continuing Exercise 4.38, let $\theta'_A(n,r) = N'_A(n,r,2)$. For all n,

$$\theta'_A(n,r) = \frac{r}{(n,r)_A} \phi_A((n,r)_A) .$$

Can this be proved using Exercise 4.8? $\theta'_A(n,r) \neq 0$ for all n . This implies that for all $s > 1$, $N'_A(n,r,s) \neq 0$ for all n .

4.40. Let A be a regular arithmetical convolution and define the arithmetical function λ_A by requiring that it be multiplicative and that for every prime p and all $\alpha \geq 1$,

$$\lambda_A(p^\alpha) = \lambda(p^{\alpha/t}) , \text{ where } t = \tau_A(p^\alpha) .$$

Define the arithmetical function β_A by

$$\beta_A(r) = \sum_{d \in A(r)} d\lambda_A(r/d) \quad \text{for all } r .$$

If $P_A(n,r,s)$ is the number of solutions $\langle x_1, \ldots, x_s \rangle$ of (*) such that $(x_i,r)_A$ is a square for $i = 1, \ldots, s$, then

$$P_A(n,r,s) = \frac{1}{r} \sum_{d \in A(r)} (\beta_A(d)\lambda_A(r/d))^s \, c_A(n,d) .$$

(Hint: use Exercises 3.12 and 4.8.)

4.41. A <u>basic sequence</u> is a set B of ordered pairs $\langle a,b \rangle$ of positive integers such that

(i) $\langle a,b \rangle \in B$ implies that $\langle b,a \rangle \in B$,

(ii) $\langle a,bc \rangle \in B$ if and only if $\langle a,b \rangle$ and $\langle a,c \rangle$ are in B ,

(iii) $\langle 1,a \rangle \in B$ for all positive integers a. Both $L = \{\langle a,b \rangle :$ a and b are positive integers$\}$ and $M = \{\langle a,b \rangle :$ a and b are positive integers and $(a,b) = 1\}$ are basic sequences. If B is a basic sequence and if $a = p_1^{\alpha_1} \ldots p_s^{\alpha_s}$ and $b = q_1^{\beta_1} \ldots q_t^{\beta_t}$, where each p_i and q_j is a prime, then $\langle a,b \rangle \in B$ if and only if $\langle p_i,q_j \rangle \in B$ for $i = 1, \ldots, s$ and $j = 1, \ldots, t$. Thus, a basic sequence B is completely determined by which pairs of primes $\langle p,q \rangle$ belong to B.

4.42. Let B be a basic sequence. If f and g are arithmetical functions, their <u>B-convolution</u> $f *_B g$ is defined by

$$(f *_B g)(n) = \sum_{\substack{d|n \\ \langle d,n/d \rangle \in B}} f(d)g(n/d) \quad \text{for all } n .$$

With respect to addition and B-multiplication the set of all arithmetical functions is a commutative ring with δ as unity. An arithmetical function f has an inverse with respect to B-multiplication if and only if $f(1) \neq 0$.

4.43. Let B be a basic sequence. An arithmetical function f is said to be <u>B-multiplicative</u> if $f(1) \neq 0$ and $f(mn) = f(m)f(n)$ whenever $\langle m,n \rangle \in B$. Thus, f is M-multiplicative if and only if it is multiplicative, and L-multiplicative if and only if it is completely multiplicative. A function f is B-multiplicative if and only if

$$f(g *_B h) = fg *_B fh$$

for all arithmetical functions f and g.

4.44. Let B be a basic sequence. If the arithmetical function

ζ_B is defined by

$$\zeta_B(n) = \begin{cases} 1 & \text{if } \langle n,n \rangle \in B \\ 0 & \text{otherwise :} \end{cases}$$

then ζ_B is B-multiplicative. If p and q are distinct primes and $A = \{\langle 1,n \rangle, \langle n,1 \rangle : n \text{ is a positive integer}\} \cup \{\langle p,p \rangle, \langle q,q \rangle\}$, then $\zeta_A *_A \zeta_A$ is not A-multiplicative. Thus, in general, the B-convolution of two B-multiplicative functions is not necessarily B-multiplicative.

4.45. Let B be a basic sequence, and let μ_B be the inverse of ζ_B with respect to B-convolution. Then $\mu_B(1) = 1$, and if $n = p_1^{\alpha_1} \ldots p_t^{\alpha_t}$,

$$\mu_B(n) = \begin{cases} (-1)^t & \text{if } \alpha_1 = \ldots = \alpha_t = 1 \text{ and } \langle p_i, p_j \rangle \in B \text{ for } 1 \leq i, j \leq t \\ 0 & \text{otherwise .} \end{cases}$$

If A is the basic sequence defined in Exercise 4.44, then μ_A is not A-multiplicative. Thus, in general, the inverse of a B-multiplicative function with respect to B-convolution is not necessarily B-multiplicative.

4.46. If B is a basic sequence such that $B \subseteq M$ then the set of B-multiplicative functions is a group with respect to B-convolution.

4.47. Let B be a basic sequence. If g_1, g_2, h_1 and h_2 are B-multiplicative functions then

$$g_1 g_2 *_B g_1 h_2 *_B h_1 g_2 *_B h_1 h_2 = (g_1 *_B h_1)(g_2 *_B h_2) *_B u ,$$

where, for all n,

$$u(n) = \begin{cases} g_1 g_2 h_1 h_2(n^{\frac{1}{2}}) & \text{if } n \text{ is a square and } \langle n^{\frac{1}{2}}, n^{\frac{1}{2}} \rangle \in B \\ 0 & \text{otherwise} . \end{cases}$$

(See Exercise 1.63.)

Notes on Chapter 4

A. A. Gioia and M. V. Subbarao [62] introduced the notion of K-convolution in the special case in which $K(n,d)$ depends only on the greatest common divisor $(d, n/d)$, i.e., in which there is an arithmetical function K' such that $K(n,d) = K'((d, n/d))$ for all pairs $\langle n,d \rangle$. If $K' = \zeta$ their generalized convolution is the Dirichlet convolution, and if $K' = \delta$ it is the unitary convolution studied by E. Cohen [60c]. They assumed that K' is a multiplicative function and that is satisfies a condition which implies that (3) holds for K, and they observed that an inversion formula holds with respect to their convolution. A. A. Gioia [65] characterized those arithmetical functions K' for which the associated K-convolution is associative, and for which inverses and convolutions of multiplicative functions are multiplicative.

K-convolutions, in the generality in which they are discussed in this chapter, were defined by T. M. K. Davison [66]. He obtained the results involving the conditions (1)-(3), and the statement in

Exercise 4.1 was proved by M. D. Gessley [67]. Extensive studies of K-convolutions and the associated rings of arithmetical functions were made by I. P. Fantino [75] and M. Ferrero [80], and an even more general convolution was considered by M. L. Fredman [70].

In both the abstract by A. A. Gioia and M. V. Subbarao [62] and the paper by T. M. K. Davison [66], the authors constructed analogues of the Ramanujan sums relative to their generalized convolutions. E. Cohen [60c] had done this earlier with respect to the unitary convolution. It should be mentioned that the unitary convolution occurred in the work of R. Vaidyanathaswamy [31], where unitary divisors were called block divisors (see Chapter 1).

Regular arithmetical convolutions were introduced by W. Narkiewicz [63]. He proved Theorem 4.1 and Corollary 4.2, and he gave the example of distinct regular arithmetical convolutions with the same primitive integers. Lemma 4.3 and the definitions of ϕ_A and $c_A(n,r)$ occur in a paper by P. J. McCarthy [68]. He observed that $c_A(\cdot,r)$ is an even function (mod r) and determined its Fourier coefficients, and he proved the results in Exercises 4.7 and 4.10-4.16.

P. J. McCarthy [71] pointed out the relation between $c_A(n,r)$ and the extended Ramanujan sum given in Exercise 4.8, and used it to prove the identity in Exercise 4.9 and to obtain the formula for $N_A(n,r,s)$ in Exercise 4.35. The special case of this formula for $A = U$ had been found earlier by E. Cohen [61k]. The formula for $N_A(n,r,s)$ in Theorem 4.4 is also due to P. J. McCarthy [77]. The unitary analogue of the Nagell function was studied by J. Morgado [62], [63a].

The A-multiplicative functions in Exercise 4.18 were defined by K. L. Yocum [73], and he gave the characterizations of these functions contained in Exercises 4.18-4.21. The properties of A-multiplicative and A-specially multiplicative functions which make up the content of Exercises 4.22-4.25 and 4.27-4.32 generalize properties of completely multiplicative and specially multiplicative functions contained in Chapter 1. These results are new, although the special case of Exercise 4.22 for $A = U$ was proved by M. V. Subbarao [68a].

The generalization of R. Vaidyanathaswamy's identical equation (see Exercise 1.67) stated in Exercise 4.26 was proved by M. V. Subbarao and A. A. Gioia [67] for the special case in which K is defined in terms of K' as in the first paragraph of these notes. Another proof was given by K. Krishna [79].

D. L. Goldsmith [68] defined basic sequences and began the study of the convolution of arithmetical functions with respect to basic sequences. In other papers on the same subject [69], [71], he obtained the results in Exercises 4.41-4.47.

A survey article on the various kinds of convolutions of arithmetical functions was written by M. V. Subbarao [72].

Chapter 5
Dirichlet Series and Generating Functions

A series of the form

$$(*) \qquad \sum_{n=1}^{\infty} \frac{f(n)}{n^s} ,$$

where f is an arithmetical function and s is a real variable, is called a Dirichlet series. It will be called <u>the Dirichlet series of f</u>. There exist Dirichlet series such that for all values of s, the series does not converge absolutely (see Exercise 5.1). If the Dirichlet series of f does converge absolutely for some values of s then for those values of s the series determines a function which, as we shall see, serves as a generating function of f.

Suppose that the Dirichlet series (*) converges absolutely for some real number s_0. If $s \geq s_0$ then

$$\left|\frac{f(n)}{n^s}\right| = \frac{|f(n)|}{n^s} \leq \frac{|f(n)|}{n^{s_0}} = \left|\frac{f(n)}{n^{s_0}}\right|$$

for all n. Therefore, by the comparison test, the series converges absolutely for all $s \geq s_0$. Let

$$s_a = \inf \{ s_0 : \text{the series } (*) \text{ converges absolutely for } s = s_0 \}.$$

It can happen that $s_a = \infty$ or $s_a = -\infty$ (see Exercise 5.1). If $-\infty \leq s_a < \infty$ then the series (*) converges absolutely for $s > s_a$.

We call s_a the <u>abscissa of absolute convergence</u> of the Dirichlet series.

If $s > s_a$ then any rearrangement of the terms of the series (*) results in a series that converges absolutely to the same value as before the rearrangement.

The series

$$\sum_{n=1}^{\infty} \frac{1}{n^s} ,$$

which is the Dirichlet series of the arithmetical function ζ , converges for $s > 1$ and diverges for $s \leq 1$: hence $s_a = 1$. The function of s that it defines for $s > 1$ is also denoted by ζ and it is called the <u>Riemann zeta function</u>. Thus,

$$\zeta(s) = \sum_{n=1}^{\infty} \frac{1}{n^s} \quad \text{for} \quad s > 1 .$$

Suppose that $s_a < \infty$ for the Dirichlet series of the arithmetical function f , and define the function F of the real variable s by

$$F(s) = \sum_{n=1}^{\infty} \frac{f(n)}{n^s} \quad \text{for} \quad s > s_a .$$

The function F uniquely determines the arithmetical function f.

<u>Proposition 5.1</u>. Suppose that the Dirichlet series of the arithmetical functions f and g converge absolutely for all $s > s_0$ and that

$$\sum_{n=1}^{\infty} \frac{f(n)}{n^s} = \sum_{n=1}^{\infty} \frac{g(n)}{n^s} \quad \text{for } s > s_0 .$$

Then $f = g$.

Proof. If $h(n) = f(n) - g(n)$ for all n then

$$\sum_{n=1}^{\infty} \frac{h(n)}{n^s} = 0 \quad \text{for } s > s_0 :$$

we shall show that $h(n) = 0$ for all n . If not, then there is a least integer m such that $h(m) \neq 0$. Then

$$h(m) = - m^s \sum_{n=m+1}^{\infty} \frac{h(n)}{n^s} .$$

The series $\sum_{n=1}^{\infty} |h(n)|/n^s$ converges to a real number R when $s = s_0 + 1$. Hence, for all $s \geq s_0 + 1$,

$$\left| \sum_{n=m+1}^{\infty} \frac{h(n)}{n^s} \right| \leq \sum_{n=m+1}^{\infty} \frac{|h(n)|}{n^{s_0+1}} \frac{1}{n^{s-s_0-1}}$$

$$\leq R \frac{1}{(m+1)^{s-s_0-1}}$$

Therefore,

$$|h(m)| \le \left(\frac{m}{m+1}\right)^s (m+1)^{s_0+1} R \to 0 \quad \text{as} \quad s \to \infty .$$

Thus, $h(m) = 0$, a contradiction. □

The function F is called the generating function of f . If G is the generating function of the arithmetical function g then $F + G$ is the generating function of $f + g$. It is easy to determine the generating function of $f * g$.

Theorem 5.2. If

$$F(s) = \sum_{n=1}^{\infty} \frac{f(n)}{n^s} \quad \text{and} \quad G(s) = \sum_{n=1}^{\infty} \frac{g(n)}{n^s} ,$$

where the Dirichlet series converge absolutely for $s > s_0$, then

$$F(s)G(s) = \sum_{n=1}^{\infty} \frac{(f * g)(n)}{n^s} \quad \text{for} \quad s > s_0 ,$$

and this series also converges absolutely for $s > s_0$.

Proof. The product of the two series converges absolutely for $s > s_0$: hence the terms of the product series can be rearranged. Thus, for $s > s_0$,

$$F(s)G(s) = \sum_{h=1}^{\infty} \sum_{k=1}^{\infty} \frac{f(h)g(k)}{(hk)^s}$$

$$= \sum_{n=1}^{\infty} \left(\sum_{hk=n} \frac{f(h)g(k)}{(hk)^s} \right)$$

$$= \sum_{n=1}^{\infty} \left(\sum_{d|n} f(d)g(n/d) \right) \frac{1}{n^s} \ . \ \square$$

For example, the Dirichlet series of the Möbius function μ converges absolutely for $s > 1$ (see Exercise 5.2), and so for $s > 1$,

$$\zeta(s) \sum_{n=1}^{\infty} \frac{\mu(n)}{n^s} = \sum_{n=1}^{\infty} \frac{(\zeta * \mu)(n)}{n^s}$$

$$= \sum_{n=1}^{\infty} \frac{\delta(n)}{n^s} = 1 \ .$$

Therefore,

$$\sum_{n=1}^{\infty} \frac{\mu(n)}{n^s} = \frac{1}{\zeta(s)} \quad \text{for} \quad s > 1 \ .$$

In fact, in general, if

$$F(s) = \sum_{n=1}^{\infty} \frac{f(n)}{n^s} \quad \text{for} \quad s > s_0 \ ,$$

and the series converges absolutely for $s > s_0$, and if f^{-1} exists, then

$$\frac{1}{F(s)} = \sum_{n=1}^{\infty} \frac{f^{-1}(n)}{n^s} \quad \text{for } s > s_0.$$

The Dirichlet series of the function ζ_1 is

$$\sum_{n=1}^{\infty} \frac{\zeta_1(n)}{n^s} = \sum_{n=1}^{\infty} \frac{1}{n^{s-1}} = \zeta(s - 1) \quad \text{for } s > 2.$$

Since $\sigma = \zeta_1 * \zeta$,

$$\sum_{n=1}^{\infty} \frac{\sigma(n)}{n^s} = \sum_{n=1}^{\infty} \frac{(\zeta_1 * \zeta)(n)}{n^s} = \zeta(s - 1)\zeta(s) \quad \text{for } s > 2.$$

Likewise, since $\phi = \zeta_1 * \mu$,

$$\sum_{n=1}^{\infty} \frac{\phi(n)}{n^s} = \frac{\zeta(s - 1)}{\zeta(s)} \quad \text{for } s > 2.$$

The generating functions of many other arithmetical functions are given throughout this chapter and in its exercises.

Let f be an arithmetical function such that the Dirichlet series of f converges absolutely to $F(s)$ for $s > s_0$. Then, for each prime p, the series

$$1 + \frac{f(p)}{p^s} + \frac{f(p^2)}{p^{2s}} + \ldots$$

converges absolutely for $s > s_0$ because it consists of some of the terms of the Dirichlet series of f. If f is a multiplicative function then $F(s)$ can be written as the infinite product [1)] of these series for all primes p. The product is called the Euler product of the Dirichlet series of f.

Theorem 5.3. Let f be a multiplicative function and suppose that the Dirichlet series of f converges absolutely for $s > s_0$. Then, for $s > s_0$,

$$\sum_{n=1}^{\infty} \frac{f(n)}{n^s} = \prod_p \left(1 + \frac{f(p)}{p^s} + \frac{f(p^2)}{p^{2s}} + \dots\right),$$

where the product is over all primes p, and the infinite product converges absolutely.

Proof. Let m be a positive integer and consider the primes $p_1,\dots,p_t$ not exceeding m. Then, for $s > s_0$,

$$\prod_{p\le m} \left(1 + \frac{f(p)}{p^s} + \frac{f(p^2)}{p^{2s}} + \dots\right)$$

$$= \sum_{j_1,\dots,j_t=0}^{\infty} \frac{f(p_1^{j_1}) \dots f(p_t^{j_t})}{(p_1^{j_1} \dots p_t^{j_t})^s} = \sum{}' \frac{f(n)}{n^s} ,$$

---

1) See [R] Chapter 1, or [Ah] Section 2.2 of Chapter 5.

where the primed sum is over all positive integers divisible only by primes from among $p_1,\ldots,p_t$. Note that we have used the assumption that f is multiplicative, and also the assumption that the Dirichlet series of f converges absolutely for $s > s_0$. Thus

$$\left| \sum_{n=1}^{\infty} \frac{f(n)}{n^s} - \prod_{p\leq m} \left(1 + \frac{f(p)}{p^s} + \frac{f(p^2)}{p^{2s}} + \ldots\right)\right| \leq \sum_{n\geq m} \frac{|f(n)|}{n^s} ,$$

and for $s > s_0$,

$$\lim_{m\to\infty} \sum_{n\geq m} \frac{|f(n)|}{n^s} = 0 .$$

Therefore, the infinite product converges to the value of the Dirichlet series.

Since

$$\sum_{p\leq m} \left|\frac{f(p)}{p^s} + \frac{f(p^2)}{p^{2s}} + \ldots\right|$$

$$\leq \sum_{p\leq m} \left(\frac{|f(p)|}{p^s} + \frac{|f(p^2)|}{p^{2s}} + \ldots\right)$$

$$\leq \sum_{n=2}^{\infty} \frac{|f(n)|}{n^s} ,$$

and since the final series converges for $s > s_0$, the partial sums of the series of nonnegative terms

$$\sum_p \left| \frac{f(p)}{p^s} + \frac{f(p^2)}{p^{2s}} + \dots \right|$$

are bounded: hence the series converges. This implies that the infinite product converges absolutely for $s > s_0$. □

Example. $f = \zeta$. For $s > 1$,

$$1 + \frac{\zeta(p)}{p^s} + \frac{\zeta(p^2)}{p^{2s}} + \dots$$

$$= 1 + \frac{1}{p^s} + \left(\frac{1}{p^s}\right)^2 + \dots = \frac{1}{1 - \frac{1}{p^s}} .$$

Thus, for $s > 1$,

$$\zeta(s) = \prod_p \frac{1}{1 - p^{-s}} .$$

Note that if r is a positive integer then for the function f of Theorem 5.3,

$$\sum_{\substack{n=1 \\ (n,r)=1}}^{\infty} \frac{f(n)}{n^s} = \prod_{p \nmid r} \left(1 + \frac{f(p)}{p^s} + \frac{f(p^2)}{p^{2s}} + \dots\right) \quad \text{for } s > s_0,$$

and that this is proved in the same way as the theorem. For example, for $s > 1$,

$$\sum_{\substack{n=1 \\ n \text{ odd}}}^{\infty} \frac{1}{n^s} = \prod_{p>2} \frac{1}{1 - p^{-s}} = (1 - 2^{-s})\zeta(s).$$

Example $f = \sigma$. For $s > 2$,

$$1 + \frac{\sigma(p)}{p^s} + \frac{\sigma(p^2)}{p^{2s}} + \dots$$

$$= 1 + \frac{p^2-1}{p^s(p-1)} + \frac{p^3-1}{p^{2s}(p-1)} + \dots$$

$$= \frac{1}{p - 1} (p(1 + p^{1-s} + p^{2(1-s)} + \dots) - (1 + p^{-s} + p^{-2s} + \dots))$$

$$= \frac{1}{p - 1} \left(\frac{p}{1 - p^{1-s}} - \frac{1}{1 - p^{-s}} \right) = \frac{1}{(1 - p^{1-s})(1 - p^{-s})}.$$

Thus, for $s > 2$,

$$\sum_{n=1}^{\infty} \frac{\sigma(n)}{n^s} = \left(\prod_p \frac{1}{1 - p^{1-s}} \right) \left(\prod_p \frac{1}{1 - p^{-s}} \right) = \zeta(s - 1)\zeta(s).$$

Note that for $s > 2$,

$$\sum_{n=1}^{\infty} \frac{\sigma(n)}{n^s} = \prod_p \frac{1}{1 - \sigma(p)p^{-s} + B(p)p^{-2s}} ,$$

where B is the completely multiplicative function associated with the specially multiplicative function σ.

There is a correspondence between the factors in the Euler product of the Dirichlet series of a multiplicative function f and the Bell series of f relative to the primes (see Exercises 1.98-1.102). The correspondence is established by means of the change of variable $X = p^{-s}$. It follows from Exercise 1.101 that if the function f of Theorem 5.3 is completely multiplicative, then

$$\sum_{n=1}^{\infty} \frac{f(n)}{n^s} = \prod_p \frac{1}{1 - f(p)p^{-s}} \quad \text{for} \quad s > s_0,$$

and if f is specially multiplicative then

$$\sum_{n=1}^{\infty} \frac{f(n)}{n^s} = \prod_p \frac{1}{1 - f(p)p^{-s} + B(p)p^{-2s}} \quad \text{for} \quad s > s_0,$$

where $B = B_f$.

Example. $f = \phi$. For $s > 2$,

$$1 + \frac{\phi(p)}{p^s} + \frac{\phi(p^2)}{p^{2s}} + \ldots$$

$$= 1 + \frac{p(1 - \frac{1}{p})}{p^s} + \frac{p^2(1 - \frac{1}{p})}{p^{2s}} + \ldots$$

$$= \frac{1}{p} + \left(1 - \frac{1}{p}\right) + \frac{1 - \frac{1}{p}}{p^{s-1}} + \frac{1 - \frac{1}{p}}{p^{2(s-1)}} + \cdots$$

$$= \frac{1}{p} + \left(1 - \frac{1}{p}\right)\left(1 + p^{1-s} + p^{2(1 - s)} + \cdots\right)$$

$$= \frac{1}{p} + \left(1 - \frac{1}{p}\right)\frac{1}{1 - p^{1-s}} = \frac{1 - p^{-s}}{1 - p^{1-s}} .$$

Therefore, for $s > 2$,

$$\sum_{n=1}^{\infty} \frac{\phi(n)}{n^s} = \prod_p \frac{1 - p^{-s}}{1 - p^{1-s}} = \frac{\zeta(s - 1)}{\zeta(s)} .$$

In the next example the function f is not necessarily multiplicative, but we shall be able to use Theorem 5.3 indirectly to obtain its generating function. In order to express the generating function neatly it will be convenient to have available the arithmetical function ϕ_s, defined for a real number s by

$$\phi_s(r) = \sum_{d|r} d^s\mu(r/d) \quad \text{for all } r.$$

If s is a positive integer then $\phi_s = J_s$. The function ϕ_s is multiplicative, and for all r,

$$\phi_s(r) = r^s \prod_{p|r} (1 - p^{-s}).$$

For a fixed r, $\phi_s(r)$ can be considered to be a function of the real variable s.

Example. Let r be a squarefree positive integer and let f be the arithmetical function defined by $f(n) = \mu(nr)$ for all n. Then f is multiplicative if and only if f has an even number of prime divisors. (In any case, $\mu(nr)/\mu(r)$ is a multiplicative function of n, by Exercise 1.11.)

$$\sum_{n=1}^{\infty} \frac{\mu(nr)}{n^s} = \mu(r) \sum_{\substack{n=1 \\ (n,r)=1}}^{\infty} \frac{\mu(n)}{n^s},$$

because $\mu(nr) = 0$ whenever $(n,r) \neq 1$, and

$$\frac{1}{\zeta(s)} = \sum_{n=1}^{\infty} \frac{\mu(n)}{n^s} = \prod_{p} (1 - p^{-s})$$

$$= \left(\prod_{p|r} (1 - p^{-s})\right)\left(\prod_{p\nmid r} (1 - p^{-s})\right) = \frac{\phi_s(r)}{r^s} \sum_{\substack{n=1 \\ (n,r)=1}}^{\infty} \frac{\mu(n)}{n^s}.$$

Therefore, for $s > 1$,

$$\sum_{n=1}^{\infty} \frac{\mu(nr)}{n^s} = \frac{\mu(r)r^s}{\phi_s(r)\zeta(s)}.$$

For our final example we shall prove a result due to S. Ramanujan.

Proposition 5.4. If h and k are positive integers then for $s > h + k + 1$,

$$\sum_{n=1}^{\infty} \frac{\sigma_h(n)\sigma_k(n)}{n^s} = \frac{\zeta(s)\zeta(s-h)\zeta(s-k)\zeta(s-h-k)}{\zeta(2s-h-k)}.$$

Proof. Suppose $s > h + k + 1$. For a prime p,

$$\sum_{m=0}^{\infty} \frac{\sigma_h(p^m)\sigma_k(p^m)}{p^{ms}} = \sum_{m=0}^{\infty} \frac{1}{p^{ms}} \frac{p^{(m+1)h} - 1}{p^h - 1} \frac{p^{(m+1)k} - 1}{p^k - 1}$$

$$= \frac{1}{(p^h - 1)(p^k - 1)} \sum_{m=0}^{\infty} \left(\frac{p^{h+k}}{p^{m(s-h-k)}} - \frac{p^h}{p^{m(s-h)}} - \frac{p^k}{p^{m(s-k)}} + \frac{1}{p^{ms}} \right)$$

$$= \frac{1}{(p^h - 1)(p^k - 1)} \left(\frac{p^{h+k}}{1 - p^{h+k-s}} - \frac{p^h}{1 - p^{h-s}} - \frac{p^k}{1 - p^{k-s}} + \frac{1}{1 - p^{-s}} \right).$$

If the four terms in the parentheses are put over the common denominator

$$(1 - p^{-s})(1 - p^{h-s})(1 - p^{k-s})(1 - p^{h+k-s}),$$

the numerator is

$$(p^h - 1)(p^k - 1)(1 - p^{h+k-2s}).$$

Thus, the sum is equal to

$$\frac{1 - p^{h+k-2s}}{(1 - p^{-s})(1 - p^{h-s})(1 - p^{k-s})(1 - p^{h+k-s})} .$$

Therefore,

$$\sum_{n=1}^{\infty} \frac{\sigma_h(n)\sigma_k(n)}{n^s}$$

$$= \prod_p \left(\frac{1}{1 - p^{-s}}\right)\left(\frac{1}{1 - p^{-(s-h)}}\right)\left(\frac{1}{1 - p^{-(s-k)}}\right)\left(\frac{1}{1 - p^{-(s-h-k)}}\right)\left(\frac{1}{1 - p^{-(2s-h-k)}}\right)^{-1},$$

which is equal to the right-hand side of the formula in the proposition. □

The same result holds when $h = 0$ and/or $k = 0$ (see Exercises 5.32 and 5.33).

It is clear from the examples and from some of the exercises that Theorem 5.3 can be used to determine, in a straight-forward manner, the generating functions of many arithmetical functions. Then Proposition 5.1 and Theorem 5.2 can be used to discover arithmetical identities involving these functions.

For example, let h and k be nonnegative integers. By Exercises 5.4 and 5.5,

$$\sum_{n=1}^{\infty} \frac{(\zeta_h * \sigma_k)(n)}{n^s} = \zeta(s - h)\zeta(s)\zeta(s - k)$$

$$= \zeta(s - k)\zeta(s)\zeta(s - h)$$

$$= \sum_{n=1}^{\infty} \frac{(\zeta_k * \sigma_h)(n)}{n^s}$$

for $s > \max(h + 1, k + 1)$. Therefore,

$$\sum_{d|n} d^h \sigma_k(n/d) = \sum_{d|n} d^k \sigma_h(n/d) \quad \text{for all } n$$

(see Exercise 1.48).

As another example, by Exercises 5.5, 5.11 and 5.33,

$$\sum_{n=1}^{\infty} \left(\sum_{d|n} \tau^2(d)\lambda(n/d)\theta(n/d) \right) \frac{1}{n^s} = \frac{\zeta(s)^4}{\zeta(2s)} \frac{\zeta(2s)}{\zeta(s)^2}$$

$$= \zeta(s)^2 = \sum_{n=1}^{\infty} \frac{\tau(n)}{n^s}$$

for $s > 1$. Therefore,

$$\sum_{d|n} \tau^2(d)\lambda(n/d)\theta(n/d) = \tau(n) \quad \text{for all } n.$$

It is easy to find the generating function of the convolution $f * g$ of two arithmetical functions f and g whose generating functions are known. But it is not so easy to find the generating function of their product fg. We shall do so now in the case when f and g are specially multiplicative, and obtain Proposition 5.4 as a special case. Another result along this line is contained in Exercise 5.44.

First, we note that if f is an arithmetical function whose Dirichlet series converges absolutely to $F(s)$ for $s > s_0$, then

$$\sum_{n=1}^{\infty} \frac{f(n)\zeta_k(n)}{n^s} = \sum_{n=1}^{\infty} \frac{f(n)}{n^{s-k}}$$

$$= F(s - k) \quad \text{for} \quad s > s_0 + k.$$

Now let f_1 and f_2 be specially multiplicative functions and let $f_1 = g_1 * h_1$ and $f_2 = g_2 * h_2$, where g_1, h_1, g_2 and h_2 are completely multiplicative. For $i = 1,2$, let F_i be the generating function of f_i, and let E_0, E_1, E_2, E_3, E_4 and U be the generating functions of $g_1h_1g_2h_2$, g_1g_2, g_1h_2, h_1g_2, h_1h_2 and u, respectively, where

$$u(n) = \begin{cases} g_1h_1g_2h_2(n^{1/2}) & \text{if } n \text{ is a square} \\ 0 & \text{otherwise.} \end{cases}$$

By Exercise 5.3, $U(s) = E_0(2s)$. If the Dirichlet series of all of these arithmetical functions converge absolutely for $s > s_0$ then by Exercise 1.63 and Theorem 5.2,

$$\sum_{n=1}^{\infty} \frac{f_1(n)f_2(n)}{n^s} = \frac{E_1(s)E_2(s)E_3(s)E_4(s)}{E_0(2s)} \quad \text{for} \quad s > s_0.$$

This reduces the problem of determining the generating function of the product of two specially multiplicative functions to one of determining the generating functions of completely multiplicative functions.

Consider the special case in which $f_1 = f$, $g_1 = g$, $h_1 = h$, $f_2 = \sigma_k$, $g_2 = \zeta$ and $h_2 = \zeta_k$. Let F, G, H and E be the generating functions of f, g, h and gh, respectively. Then

$$g_1g_2 = g, \quad g_1h_2 = g\zeta_k, \quad h_1g_2 = h, \quad h_1h_2 = h\zeta_k$$

and

$$E_1(s) = G(s), \quad E_2(s) = G(s - k), \quad E_3(s) = H(s), \quad E_4(s) = H(s - k).$$

Thus

$$E_1(s)E_3(s) = G(s)H(s) = F(s)$$

and

$$E_2(s)E_4(s) = G(s - k)H(s - k) = F(s - k).$$

Also, $E_0(2s) = E(2s - k)$, and therefore

$$\sum_{n=1}^{\infty} \frac{f(n)\sigma_k(n)}{n^s} = \frac{F(s)F(s - k)}{E(2s - k)} \quad \text{for } s > s_0.$$

If $f = \sigma_h$ we obtain the result of Proposition 5.4.

For a real number s let σ_s be the arithmetical function defined by

$$\sigma_s(n) = \sum_{d|n} d^s \quad \text{for all } n.$$

Then, for a fixed n, $\sigma_s(n)$ is a function of the real variable s. This function arises in the consideration of the Dirichlet series of certain arithmetical functions obtained from the Ramanujan sums.

<u>Theorem 5.5.</u> If $s > 1$ then for all n,

$$\frac{\sigma_{s-1}(n)}{n^{s-1}} = \zeta(s) \sum_{r=1}^{\infty} \frac{c(n,r)}{r^s} .$$

This result can be viewed in two ways. First of all, it states that

$$\frac{\sigma_{s-1}(n)}{\zeta(s)n^{s-1}}$$

is the generating function of the arithmetical function $c(n,\cdot)$. On the other hand, it gives a representation of the arithmetical function σ_{s-1} as a series of terms involving Ramanujan sums. We will have more to say about this after we prove the theorem.

Proof of Theorem 5.5. Let n be fixed. Then

$$|c(n,r)| \le \sum_{d|(n,r)} d \le \sum_{d|n} d = \sigma(n) \quad \text{for all} \quad r:$$

hence the Dirichlet series

$$\sum_{r=1}^{\infty} \frac{c(n,r)}{r^s}$$

converges absolutely for $s > 1$ by Exercise 5.2. Thus, for $s > 1$, by Theorem 5.2,

$$\zeta(s) \sum_{r=1}^{\infty} \frac{c(n,r)}{r^s} = \sum_{r=1}^{\infty} \left(\sum_{d|r} c(n,d) \right) \frac{1}{r^s}$$

and (see the proof of Proposition 2.1) this is equal to

$$\sum_{r|n} \frac{1}{r^{s-1}} = \frac{1}{n^{s-1}} \sum_{r|n} \left(\frac{n}{r}\right)^{s-1} = \frac{\sigma_{s-1}(n)}{n^{s-1}} . \quad \square$$

If k is a positive integer then for all n,

$$\sigma_k(n) = \zeta(k+1)n^k \sum_{r=1}^{\infty} \frac{c(n,r)}{r^{k+1}} .$$

The exact value of $\zeta(k+1)$ is known when k is odd. In fact,

$$\zeta(2h) = (-1)^{h+1} \frac{(2\pi)^{2h} B_{2h}}{2(2h)!} ,$$

where B_{2h} is the (2h)th Bernoulli number.[1] In particular, $\zeta(2) = \pi^2/6$ and

$$\sigma(n) = \frac{1}{6}\pi^2 n \sum_{r=1}^{\infty} \frac{c(n,r)}{r^2} \quad \text{for all } n.$$

Note that

$$\frac{\sigma_{s-1}(n)}{n^{s-1}} = \sum_{d|n} \frac{1}{d^{s-1}} = \sum_{r=1}^{\infty} \frac{f(r)}{r^s} ,$$

where

---

1) See [A] p. 266. $B_2 = \frac{1}{6}$, $B_4 = -\frac{1}{30}$, $B_6 = \frac{1}{42}$, $B_8 = -\frac{1}{30}$, $B_{10} = \frac{5}{66}$, etc.

$$f(r) = \begin{cases} r & \text{if } r \mid n \\ 0 & \text{if } r \nmid n , \end{cases}$$

and that this Dirichlet series converges (absolutely) for all s because it has only a finite number of nonzero terms. Note also that for all r,

$$(f * \mu)(r) = \sum_{d \mid (n,r)} d\mu(r/d) = c(n,r) .$$

Therefore, by Theorem 5.2, if $s > 1$,

$$\sum_{r=1}^{\infty} \frac{c(n,r)}{r^s} \quad \text{converges to} \quad \frac{\sigma_{s-1}(n)}{n^{s-1}} \sum_{r=1}^{\infty} \frac{\mu(r)}{r^s} :$$

this is simply an alternate phrasing of the proof of Theorem 5.5.

What can we say about the case in which $s = 1$? It is true that

$$\sum_{r=1}^{\infty} \frac{\mu(r)}{r} = 0 ,$$

but this is much deeper than the results discussed in this book. In fact, it is a statement equivalent to the prime number theorem.[1)] The series of terms $\mu(r)/r$ does not converge absolutely since for all m,

$$\sum_{p \leq m} \frac{1}{p} \leq \sum_{r=1}^{m} \frac{|\mu(r)|}{r} ,$$

1) See [Ay] Chap. II, especially pp. 113-116.

and it is an elementary fact that the series of the reciprocals of the primes diverges.[1)] However, the hypotheses of Theorem 5.2 can be weakened enough to allow us to conclude that for all n

$$\sum_{r=1}^{\infty} \frac{c(n,r)}{r} \quad \text{converges to} \quad \tau(n) \sum_{r=1}^{\infty} \frac{\mu(r)}{r} ,$$

i.e.,

$$\sum_{r=1}^{\infty} \frac{c(n,r)}{r} = 0 .$$

Suppose that for some fixed real number s the first series in Theorem 5.2 converges absolutely to A and the second converges (but not necessarily absolutely) to B. We claim that the third series converges to AB.

Let

$$\sum_{r=1}^{m} \frac{f(r)}{r^s} = A_m , \sum_{r=1}^{m} \frac{|f(r)|}{r^s} = A'_m , \sum_{r=1}^{m} \frac{g(r)}{r^s} = B_m .$$

There is a real number M such that $A'_m < M$ and $|B_m| < M$ for all m. Let $\varepsilon > 0$. There is a positive integer $m_0 = m_0(\varepsilon)$ such that for all $m \geq m_0$,

---

1) See [A] p. 18.

$$\sum_{r=[\sqrt{m}]+1}^{m} \frac{|f(r)|}{r^s} < \frac{\varepsilon}{3M+1} ,$$

$$|B_k - B_{m_0}| < \frac{\varepsilon}{3M+1} \quad \text{for all } k \text{ with } \sqrt{m} \le k \le m$$

and

$$|A_m B_m - AB| < \frac{\varepsilon}{3M+1} .$$

Now, for all m ,

$$\sum_{r=1}^{m} \frac{(f * g)(r)}{r^s} = \sum_{hk \le m} \frac{f(h)}{h^s} \frac{g(k)}{k^s} = \sum_{r=1}^{m} \frac{f(r)}{r^s} B_{[m/r]} ,$$

and so

$$\sum_{r=1}^{m} \frac{(f * g)(r)}{r^s} - A_m B_m = \sum_{r=1}^{m} \frac{f(r)}{r^s} (B_{[m/r]} - B_m)$$

$$= \sum_{r=1}^{[\sqrt{m}]} \frac{f(r)}{r^s} (B_{[m/r]} - B_m) + \sum_{r=[\sqrt{m}]+1}^{m} \frac{f(r)}{r^s} (B_{[m/r]} - B_m) .$$

Therefore, if $m \ge m_0$,

$$\left| \sum_{r=1}^{m} \frac{(f * g)(r)}{r^s} - A_m B_m \right|$$

$$< \frac{\varepsilon}{3M+1} \sum_{r=1}^{[\sqrt{m}]} \frac{|f(r)|}{r^s} + 2M \sum_{r=[\sqrt{m}]+1}^{m} \frac{|f(r)|}{r^s}$$

$$< \frac{3M}{3M+1} \varepsilon :$$

hence

$$\left| \sum_{r=1}^{m} \frac{(f * g)(r)}{r^s} - AB \right| < \frac{3M}{3M+1} \varepsilon + |A_m B_m - AB| < \varepsilon .$$

This completes the proof of the claim.

We can consider also the Dirichlet series of the arithmetical function $c(\cdot,r)$, i.e., the series

$$\sum_{n=1}^{\infty} \frac{c(n,r)}{n^s} .$$

Since $|c(n,r)| \leq \sigma(r)$ for all $n \geq 1$, this series converges absolutely for $s > 1$.

For $s > 1$,

$$\sum_{n=1}^{\infty} \frac{c(n,r)}{n^s} = \sum_{n=1}^{\infty} \left(\sum_{d|(n,r)} d\mu(r/d) \right) \frac{1}{n^s} .$$

For each divisor d of r there is one term for each multiple md of d: hence this is equal to

$$\sum_{d|r} \left(d\mu(r/d) \sum_{m=1}^{\infty} \frac{1}{(md)^s} \right) = \left(\sum_{m=1}^{\infty} \frac{1}{m^s} \right) \left(\sum_{d|r} d^{1-s}\mu(r/d) \right) .$$

We have shown that if $s > 1$ then for all r,

$$\sum_{n=1}^{\infty} \frac{c(n,r)}{n^s} = \zeta(s)\phi_{1-s}(r) .$$

The manner in which this formula has been derived can be applied in a more general situation, and this is done in Exercise 5.47.

There is a general result that has Theorem 5.5 as a special case, and has other special cases of interest as well. It is a consequence of the following inversion formula.

Proposition 5.6. Let f and g be arithmetical functions and assume that

$$\sum_{k=1}^{\infty} \sum_{m=1}^{\infty} g(kmn)$$

converges absolutely for all n. If

(A) $$f(n) = \sum_{m=1}^{\infty} g(mn) \quad \text{for all } n$$

then

(B) $$g(n) = \sum_{m=1}^{\infty} \mu(m) f(mn) \quad \text{for all } n.$$

Proof. Assume that (A) holds. Then

$$\sum_{m=1}^{\infty} \mu(m) f(mn) = \sum_{m=1}^{\infty} \sum_{r=1}^{\infty} \mu(m) g(mnr),$$

and the series on the right-hand side converges absolutely for all n. Hence, we can rearrange its terms so that it becomes

$$\sum_{r=1}^{\infty} g(nr) \sum_{m|r} \mu(m) = g(n)$$

for all n. □

Note that if we proceed formally or, what is the same in this instance, if we assume that all of the series in question converge absolutely, then (A) and (B) are equivalent. However this is not true in general. For example, let $f(n) = 1/n$ and $g(n) = 0$ for all n. Then the hypothesis of the proposition holds and for all n,

$$\sum_{m=1}^{\infty} \mu(m)f(mn) = \frac{1}{n} \sum_{m=1}^{\infty} \frac{\mu(m)}{m} = 0 = g(n),$$

as we have noted earlier in the chapter. Thus (B) holds, but clearly (A) does not.

Theorem 5.7. Let g be an arithmetical function such that

$$\sum_{k=1}^{\infty} \sum_{m=1}^{\infty} g(kmn)$$

converges absolutely for all n, and suppose that

$$f(n) = \sum_{d|n} dg(d) \quad \text{for all } n.$$

Let

$$h(n) = \sum_{m=1}^{\infty} g(mn) \quad \text{for all } n.$$

Then

$$f(n) = \sum_{r=1}^{\infty} h(r)c(n,r) \quad \text{for all} \quad n ,$$

and the series converges absolutely for all n .

Proof. By the preceding proposition,

$$g(n) = \sum_{m=1}^{\infty} \mu(m)h(mn) \quad \text{for all} \quad n ,$$

and the series converges absolutely for all n . Hence

$$f(n) = \sum_{d|n} d \sum_{m=1}^{\infty} \mu(m)h(md)$$

$$= \sum_{r=1}^{\infty} h(r) \sum_{d|(n,r)} d\mu(r/d)$$

$$= \sum_{r=1}^{\infty} h(d)c(n,r) . \square$$

<u>Example</u>. (Theorem 5.5) If $s > 1$ then for all n ,

$$\sigma_{1-s}(n) = \sum_{d|n} d^{1-s} = \sum_{d|n} d \frac{1}{d^s} ,$$

and

$$\sum_{k=1}^{\infty} \sum_{m=1}^{\infty} \frac{1}{(kmn)^s}$$

converges absolutely for all n. Thus, for all n,

$$\sigma_{1-s}(n) = \sum_{r=1}^{\infty} h(r)c(n,r)$$

where

$$h(n) = \sum_{m=1}^{\infty} \frac{1}{(mn)^s} = \frac{1}{n^s} \sum_{m=1}^{\infty} \frac{1}{m^s} = \frac{\zeta(s)}{n^s} .$$

Therefore, for all n

$$\frac{\sigma_{s-1}(n)}{n^{s-1}} = \sigma_{1-s}(n) = \zeta(s) \sum_{r=1}^{\infty} \frac{c(n,r)}{r^s} .$$

We shall use Theorem 5.7 to prove that for all n,

$$J_k(n) = \frac{n^k}{\zeta(k+1)} \sum_{r=1}^{\infty} \frac{\mu(r)}{J_{k+1}(r)} \; c(n,r) \, .$$

In particular,

$$\phi(n) = \frac{6n}{\pi^2} \sum_{r=1}^{\infty} \frac{\mu(r)}{J_2(r)} \; c(n,r) \, .$$

Theorem 5.8. If $s > 1$ then for all n,

$$\phi_{s-1}(n) = \frac{n^{s-1}}{\zeta(s)} \sum_{r=1}^{\infty} \frac{\mu(r)}{\phi_s(r)} \; c(n,r) \, .$$

Proof. Let $s > 1$. For all n,

$$\phi_{s-1}(n) = \sum_{d|n} d^{s-1}\mu(r/d) = n^{s-1} \sum_{d|n} \frac{\mu(d)}{d^{s-1}}$$

$$= n^{s-1} \sum_{d|n} d \, \frac{\mu(d)}{d^s} \, ,$$

and

$$\sum_{k=1}^{\infty} \sum_{m=1}^{\infty} \frac{\mu(kmn)}{(kmn)^s}$$

converges absolutely for all n. Hence

$$\phi_{s-1}(n) = n^{s-1} \sum_{r=1}^{\infty} h(r)c(n,r) ,$$

where

$$h(r) = \sum_{m=1}^{\infty} \frac{\mu(mr)}{(mr)^s} = \frac{1}{r^s} \sum_{m=1}^{\infty} \frac{\mu(mr)}{m^s}$$

$$= \frac{\mu(r)}{\phi_s(r)\zeta(s)} . \square$$

Next we shall consider some Dirichlet series of the type

$$\sum_{\substack{r=1 \\ (r,n)=1}}^{\infty} \frac{f(r)}{r^s} .$$

This series converges absolutely for those values of s for with the Dirichlet series of f converges absolutely. The following result is useful in the investigation of series of this type. It should be compared with Theorem 5.7.

<u>Proposition 5.9</u>. Let g be an arithmetical function such that

$$\sum_{k=1}^{\infty} \sum_{m=1}^{\infty} g(kmn)$$

converges absolutely for all n, and suppose that

$$f(n) = \sum_{d|n} \mu(d)g(d) \quad \text{for all } n .$$

If

$$h(n) = \sum_{m=1}^{\infty} \mu(m)g(mn) \quad \text{for all } n$$

then

$$f(n) = \sum_{\substack{r=1 \\ (r,n)=1}}^{\infty} h(r) \quad \text{for all } n .$$

and the series converges absolutely for all n.

Proof. By Exercise 5.61,

$$g(n) = \sum_{m=1}^{\infty} h(mn) \quad \text{for all } n .$$

Thus, for all n,

$$f(n) = \sum_{d|n} \mu(d) \sum_{m=1}^{\infty} h(md) = \sum_{d|n} \mu(d) \sum_{\substack{r=1 \\ d|r}}^{\infty} h(r)$$

$$= \sum_{r=1}^{\infty} h(r) \sum_{d|(n,r)} \mu(d) = \sum_{\substack{r=1 \\ (n,r)=1}}^{\infty} h(r) \ . \ \square$$

<u>Example</u>. Assume $s > 1$ and let

$$g(n) = \frac{1}{n^s} \quad \text{and} \quad f(n) = \frac{\phi_s(n)}{n^s} \quad \text{for all } n \ .$$

Then $\sum_{k=1}^{\infty} \sum_{m=1}^{\infty} g(kmn)$ converges absolutely for all n, and

$$h(n) = \sum_{m=1}^{\infty} \frac{\mu(m)}{(mn)^s} = \frac{1}{n^s} \sum_{m=1}^{\infty} \frac{\mu(m)}{m^s} = \frac{1}{n^s \zeta(s)} \ .$$

Therefore, for all n,

$$\frac{\phi_s(n)}{n^s} = \frac{1}{\zeta(s)} \sum_{\substack{r=1 \\ (r,n)=1}}^{\infty} \frac{1}{r^s} \qquad \text{for } s > 1 \ .$$

(This also follows from a formula found in the fourth example following Theorem 5.3, and Exercise 5.63.) For an integer $k \geq 2$,

$$J_k(n) = \frac{n^k}{\zeta(k)} \sum_{\substack{r=1 \\ (r,n)=1}}^{\infty} \frac{1}{r^k} \qquad \text{for all } n \ ,$$

and in particular,

$$J_2(n) = \frac{6n^2}{\pi^2} \sum_{\substack{r=1 \\ (r,n)=1}}^{\infty} \frac{1}{r^2} \quad \text{for all } n .$$

The convergence is fairly fast. For example, $J_2(10) = 72$ and the sum of the first ten terms on the right-hand side is approximately 70.37.

Example. For a real number s let ψ_s be the arithmetical function defined by

$$\psi_s(n) = \sum_{d|n} d^s |\mu(n/d)| \quad \text{for all } n .$$

If s is a positive integer then ψ_s is the generalization of Dedekind's function ψ defined in Exercise 1.34. For a prime p and $\alpha \geq 1$,

$$\psi_s(p^\alpha) = p^{\alpha s}\left(1 + \frac{1}{p^s}\right) .$$

Assume $s > 1$. For all n,

$$\frac{\psi_s(n)}{n^s} = \frac{1}{n^s} \sum_{d|n} \left(\frac{n}{d}\right)^s |\mu(d)| = \sum_{d|n} \mu(d) \frac{\mu(d)}{d^s} .$$

If we let $g(n) = \mu(n)/n^s$ for all n then $\sum_{k=1}^{\infty} \sum_{m=1}^{\infty} g(kmn)$ converges absolutely for all n, and

$$h(n) = \sum_{m=1}^{\infty} \mu(m) \frac{\mu(mn)}{(mn)^s}$$

$$= \frac{\mu(n)}{n^s} \sum_{\substack{m=1 \\ (m,n)=1}}^{\infty} \frac{|\mu(m)|}{m^s} \quad \text{for all } n .$$

Now for $s > 1$,

$$\sum_{\substack{m=1 \\ (m,n)=1}}^{\infty} \frac{|\mu(m)|}{m^s} = \frac{\prod_p (1 + \frac{1}{p^s})}{\prod_{p|n} (1 + \frac{1}{p^s})} \cdot \frac{\prod_p (1 - \frac{1}{p^s})}{\prod_p (1 - \frac{1}{p^s})}$$

$$= \frac{n^s \prod_p (1 - \frac{1}{p^{2s}})}{\psi_s(n)} \cdot \frac{1}{\prod_p (1 - \frac{1}{p^s})}$$

$$= \frac{n^s \zeta(s)}{\psi_s(n) \zeta(2s)} :$$

hence

$$h(n) = \frac{\mu(n)\zeta(s)}{\psi_s(n)\zeta(2s)} \quad \text{for all } n .$$

Thus, for $s > 1$,

$$\frac{\psi_s(n)}{n^s} = \frac{\zeta(s)}{\zeta(2s)} \sum_{\substack{r=1 \\ (r,n)=1}}^{\infty} \frac{\mu(r)}{\psi_s(r)} \quad \text{for all } n .$$

By Exercise 5.41,

$$\sum_{\substack{r=1 \\ (r,n)=1}}^{\infty} \frac{\mu(r)}{\psi_s(r)} = \prod_{p\nmid n} \left(1 - \frac{1}{p^s+1}\right) = \prod_{p\nmid n} \frac{1}{1+p^{-s}}$$

$$= \prod_{p\nmid n} \left(1 - \frac{1}{p^s} + \frac{1}{p^{2s}} - \frac{1}{p^{3s}} + \cdots\right)$$

$$= \prod_{p\nmid n} \left(1 + \frac{\lambda(p)}{p^s} + \frac{\lambda(p^2)}{p^{2s}} + \frac{\lambda(p^3)}{p^{3s}} + \cdots\right)$$

$$= \sum_{\substack{r=1 \\ (r,n)=1}}^{\infty} \frac{\lambda(r)}{r^s} ,$$

where λ is Liouville's function (see Exercise 1.47). Therefore, for $s > 1$,

$$\frac{\psi_s(n)}{n^s} = \frac{\zeta(s)}{\zeta(2s)} \sum_{\substack{r=1 \\ (r,n)=1}}^{\infty} \frac{\lambda(r)}{r^s} \quad \text{for all } n .$$

In particular,

$$\psi_2(n) = \frac{15n^2}{\pi^2} \sum_{\substack{r=1 \\ (r,n)=1}}^{\infty} \frac{\lambda(r)}{r^2} \quad \text{for all } n .$$

$\psi_2(10) = 130$, and the sum of the first ten terms of

$$\frac{1500}{\pi^2} \sum_{\substack{r=1 \\ (r,10)=1}}^{\infty} \frac{\lambda(r)}{r^2} = \frac{1500}{\pi^2} \left(1 - \frac{1}{3^2} - \frac{1}{7^2} - \frac{1}{11^2} - \frac{1}{13^2} - \frac{1}{17^2} - \frac{1}{19^2} + \frac{1}{21^2} - \frac{1}{23^2} - \frac{1}{27^2} + \cdots\right)$$

is approximately 128.74.

There is another simple general principle that leads to results of the kind obtained in this chapter.

Proposition 5.10. Let f be an arithmetical funciton such that $\sum_{r=1}^{\infty} f(r)$ converges absolutely. Then

$$\sum_{r=1}^{\infty} f(r) = \lim_{m\to\infty} \sum_{d|m!} f(d) .$$

Proof.

$$\left| \sum_{r=1}^{m} f(r) - \sum_{d|m!} f(d) \right| = \left| \sum_{\substack{d|m! \\ d>m}} f(d) \right|$$

$$\leq \sum_{\substack{d|m \\ d>m}} |f(d)| < \sum_{r=m+1}^{\infty} |f(r)| \to 0 \quad \text{as} \quad m \to \infty . \square$$

Example. (Theorem 5.5) We begin by noting that

$$\sum_{r=1}^{\infty} \frac{c(n,r)}{r^s}$$

converges absolutely for $s > 1$: hence

$$\sum_{r=1}^{\infty} \frac{c(n,r)}{r^s} = \lim_{m\to\infty} \sum_{d|m!} \frac{c(n,d)}{d^s} \quad \text{for} \quad s > 1 .$$

By Exercise 2.37,

$$\sum_{d|r} \frac{c(n,d)}{d^s} = \frac{1}{r^s} \sum_{d|(n,r)} d\, \phi_s(r/d)$$

$$= \frac{1}{r^s} \sum_{d|(n,r)} d \sum_{e|\frac{r}{d}} \left(\frac{r}{de}\right)^s \mu(e)$$

$$= \sum_{d|(n,r)} \frac{1}{d^{s-1}} \sum_{e|\frac{r}{d}} \frac{1}{e^s} \mu(e) .$$

(In the exercise, s is a positive integer, but the identity holds for all real s : prove this!) Thus, for $s > 1$,

$$\sum_{r=1}^{\infty} \frac{c(n,r)}{r^s} = \lim_{m\to\infty} \sum_{d|(n,m!)} \frac{1}{d^{s-1}} \sum_{e|\frac{m!}{d}} \frac{1}{e^s}\mu(e) ,$$

For fixed n , $n|m!$ when m is large. Hence this is equal to

$$\sum_{d|n} \frac{1}{d^{s-1}} \left(\lim_{m\to\infty} \sum_{e|\frac{m!}{d}} \frac{1}{e^s}\mu(e)\right)$$

$$= \left(\sum_{d|n} \frac{1}{d^{s-1}}\right)\left(\sum_{e=1}^{\infty} \frac{\mu(e)}{e^s}\right) = \frac{\sigma_{s-1}(n)}{n^{s-1}} \frac{1}{\zeta(s)} ,$$

and Theorem 5.5 follows.

Example. We can also use Proposition 5.10 to prove Theorem 5.8. This requires *a priori* determination of the absolute convergence of

$$\sum_{r=1}^{\infty} \frac{\mu(r)}{\phi_s(r)} c(n,r)$$

for $s > 1$. Assuming for the moment that this has been done, we have

$$\sum_{r=1}^{\infty} \frac{\mu(r)}{\phi_s(r)} c(n,r) = \lim_{m\to\infty} \sum_{d|m!} \frac{\mu(d)}{\phi_s(d)} c(n,d) .$$

By Exercise 2.44 (the identity there holds for any real number s),

$$\frac{\phi_{s-1}((n,r))}{(n,r)^{s-1}} = \frac{\phi_s(r)}{r^s} \sum_{d|r} \frac{\mu(d)}{\phi_s(d)} c(n,d)$$

$$= \sum_{d|r} \frac{\mu(d)}{d^s} \sum_{d|r} \frac{\mu(d)}{\phi_s(d)} c(n,d) \quad .$$

If we set $r = m!$ and then let $m \to \infty$ we obtain

$$\frac{\phi_{s-1}(n)}{n^{s-1}} = \frac{1}{\zeta(s)} \sum_{r=1}^{\infty} \frac{\mu(r)}{\phi_s(r)} c(n,r) ,$$

which is Theorem 5.8.

Since $|\mu(r)c(n,r)| \le \sigma(n)$ for all r, to show that the series in this example converges absolutely for $s > 1$ it suffices to show that if $\varepsilon > 0$ and $s - \varepsilon > 1$, there is a real number K such that

$$\frac{1}{\phi_s(r)} \le K \frac{1}{r^{s-\varepsilon}}$$

a fact which is a consequence of the second lemma proved below.

<u>Lemma 5.11</u>. If f is a multiplicative arithmetical function and if $f(p^\alpha) \to 0$ as the prime power $p^\alpha \to \infty$, then $\lim_{r\to\infty} f(r) = 0$.

Proof. By hypothesis there are real numbers $a > 1$ and b such that $|f(p^\alpha)| < a$ for all primes p and all $\alpha \ge 1$ and $|f(p^\alpha)| < 1$ for all $p^\alpha > b$. The numbers a and b do not depend on p or on α but only on the function f.

Let $\varepsilon > 0$. There is a real number c that depends on ε such that

$$|f(p^\alpha)| < \varepsilon \quad \text{for all } p^\alpha > c .$$

Let m be the number of prime powers $p^{\alpha} \le b$. If $r = p_1^{\alpha_1} \dots p_t^{\alpha_t}$ then there are at most m of the factors $p_i^{\alpha_i}$ with $p_i^{\alpha_i} \le b$, and for each such factor, $|f(p_i^{\alpha_i})| < a$. For each of the other factors $p_j^{\alpha_j}$, $|f(p_j^{\alpha_j})| < 1$. If some $p_j^{\alpha_j} > c$ then $|f(p_j^{\alpha_j})| < \varepsilon$ and

$$|f(r)| = |f(p_1^{\alpha_1})| \dots |f(p_t^{\alpha_t})| < a^m \varepsilon .$$

There is only a finite number of prime powers $p^{\alpha} \le c$, and there is only a finite number of integers that are products of such prime powers. If r_0 is greater than all such integers then

$$|f(r)| < a^m \varepsilon \quad \text{for all } r > r_0 . \ \Box$$

<u>Lemma 5.12</u>. For $s \ge 1$ and $\varepsilon > 0$,

$$\lim_{r \to \infty} \frac{r^{s-\varepsilon}}{\phi_s(r)} = 0 .$$

Proof. Let $f(r) = r^{s-\varepsilon}/\phi_s(r)$ for all r: then f is a multiplicative function. Hence, it is sufficient to show that

$$f(p^{\alpha}) \to 0 \quad \text{as} \quad p^{\alpha} \to \infty .$$

In fact,

$$f(p^{\alpha}) = \frac{p^{\alpha(s-\varepsilon)}}{p^{\alpha s}(1 - \frac{1}{p^s})} = \frac{1}{p^{\alpha\varepsilon}(1 - \frac{1}{p^s})} \le \frac{1}{\frac{1}{2}p^{\alpha\varepsilon}} \to 0 \quad \text{as} \quad p^{\alpha} \to \infty . \ \Box$$

Exercises for Chapter 5

5.1. (a) The Dirichlet series $\sum_{n=1}^{\infty} \frac{n^n}{n^s}$ diverges for all s: hence $s_a = \infty$.

(b) The Dirichlet series $\sum_{n=1}^{\infty} \frac{n^{-n}}{n^s}$ converges for all s: hence $s_a = -\infty$.

5.2. If f is a bounded arithmetical function then for the Dirichlet series of f, $s_a \le 1$.

5.3. Let f be an arithmetical function and k a positive integer. If $F(s)$ is the generating function of f for $s > s_0$ then $F(ks)$ is the generating function of $\Omega_k(f)$ for $s > s_0/k$. (See Exercise 1.83.)

5.4. $$\sum_{n=1}^{\infty} \frac{\zeta_k(n)}{n^s} = \zeta(s - k) \quad \text{for } s > k + 1.$$

5.5. $$\sum_{n=1}^{\infty} \frac{\sigma_k(n)}{n^s} = \zeta(s)\zeta(s - k) \quad \text{for } s > k + 1.$$

5.6. $$\sum_{n=1}^{\infty} \frac{J_k(n)}{n^s} = \frac{\zeta(s - k)}{\zeta(s)} \quad \text{for } s > k + 1.$$

5.7. $$\sum_{n=1}^{\infty} \frac{\lambda(n)}{n^s} = \frac{\zeta(2s)}{\zeta(s)} \quad \text{for } s > 1.$$

5.8. $\displaystyle\sum_{n=1}^{\infty} \frac{|\mu(n)|}{n^s} = \frac{\zeta(s)}{\zeta(2s)}$ for $s > 1$. (Hint: what is λ^{-1}?)

5.9. $\displaystyle{\sum}' \frac{1}{n^s} = \frac{1}{2}\left(\frac{\zeta(s)}{\zeta(2s)} - \frac{1}{\zeta(s)}\right)$ for $s > 1$ where the primed sum is over all integers n which are products of an odd number of primes. (Hint: consider the function f defined by $f(n) = |\mu(n)| - \mu(n)$ for all n.) In particular,

$$\sum{}' \frac{1}{n^2} = \frac{9}{2\pi^2}$$

5.10. $\displaystyle\sum_{n=1}^{\infty} \frac{\omega(n)}{n^s} = \zeta(s) \sum_{p} \frac{1}{p^s}$ for $s > 1$, where the sum on the right-hand side is over all primes p. (Hint: see Exercise 1.17, in which ω is defined, and examine $\mu * \omega$.)

5.11. $\displaystyle\sum_{n=1}^{\infty} \frac{\theta(n)}{n^s} = \frac{\zeta(s)^2}{\zeta(2s)}$ for $s > 1$

and

$$\sum_{n=1}^{\infty} \frac{\lambda(n)\theta(n)}{n^s} = \frac{\zeta(2s)}{\zeta(s)^2} \quad \text{for } s > 1.$$

(See Exercises 1.24 and 1.47.)

5.12. If μ_k is the function defined in Exercise 1.30 then $\mu_k^{-1} = \nu_k$, which is defined in Exercise 1.89. The Dirichlet series of these functions are

$$\sum_{n=1}^{\infty} \frac{\mu_k(n)}{n^s} = \frac{1}{\zeta(ks)} \quad \text{and} \quad \sum_{n=1}^{\infty} \frac{\nu_k(n)}{n^s} = \zeta(ks)$$

for $s > 1/k$. For Klee's function Φ_k (see Exercise 1.29),

$$\sum_{n=1}^{\infty} \frac{\Phi_k(n)}{n^s} = \frac{\zeta(s-1)}{\zeta(ks)} \quad \text{for } s > 2.$$

5.13. For a positive integer k let ξ_k be the arithmetical function defined by

$$\xi_k(n) = \begin{cases} 1 & \text{if } n \text{ is k-free} \\ 0 & \text{otherwise.} \end{cases}$$

Then

$$\sum_{n=1}^{\infty} \frac{\xi_k(n)}{n^s} = \frac{\zeta(s)}{\zeta(ks)} \quad \text{for } s > 1.$$

5.14. For the arithmetical functions τ_k and $\tau_{k,h}$ defined in Exercises 1.32 and 1.33,

$$\sum_{n=1}^{\infty} \frac{\tau_k(n)}{n^s} = \zeta(s)^k \quad \text{for } s > 1$$

and

$$\sum_{n=1}^{\infty} \frac{\tau_{k,h}(n)}{n^s} = \zeta(hs)^k \quad \text{for} \quad s > 1/h.$$

5.15. For all positive integers k and h, with $k \geq 2$,

$$\tau_{k,h}(n) = \sum \nu_h(d_1)\ldots\nu_h(d_k) \quad \text{for all} \quad n,$$

where the sum is over all ordered k-tuples $\langle d_1,\ldots,d_k\rangle$ of divisors of n. In particular, $\tau_4 = \tau * \tau$.

5.16. $$\sum_{n=1}^{\infty} \frac{\beta_k(n)}{n^s} = \frac{\zeta(s - k)\zeta(2s)}{\zeta(s)} \quad \text{for } s > k + 1, \text{ where}$$

β_k is the arithmetical function defined in Exercise 1.78. In particular,

$$\sum_{n=1}^{\infty} \frac{\beta(n)}{n^s} = \frac{\zeta(s - 1)\zeta(2s)}{\zeta(s)} \quad \text{for} \quad s > 2$$

($\beta = \beta_1$ was introduced near the end of Chapter 1).

5.17. For Gegenbauer's function $\rho_{k,t}$ (see Exercise 1.89),

$$\sum_{n=1}^{\infty} \frac{\rho_{k,t}(n)}{n^s} = \zeta(s - k)\zeta(ts) \quad \text{for} \quad s > k + 1$$

5.18. For the arithmetical function ψ_k defined in Exercise 1.34,

$$\sum_{n=1}^{\infty} \frac{\psi_k(n)}{n^s} = \frac{\zeta(s - k)\zeta(s)}{\zeta(2s)} \quad \text{for} \quad s > k + 1.$$

In particular, the Dirichlet series of Dedekind's function ψ is

$$\sum_{n=1}^{\infty} \frac{\psi(n)}{n^s} = \frac{\zeta(s - 1)\zeta(s)}{\zeta(2s)} \quad \text{for} \quad s > 2.$$

5.19. For the arithmetical functions q_k and Ψ_k defined in Exercise 1.35,

$$\sum_{n=1}^{\infty} \frac{q_k(n)}{n^s} = \frac{\zeta(ks)}{\zeta(2ks)} \quad \text{for} \quad s > 1/k$$

and

$$\sum_{n=1}^{\infty} \frac{\Psi_k(n)}{n^s} = \frac{\zeta(s - 1)\zeta(ks)}{\zeta(2ks)} \quad \text{for} \quad s > 2.$$

5.20. For the arithmetical function R defined at the end of Chapter 1,

$$\sum_{n=1}^{\infty} \frac{R(n)}{n^s} = 4\zeta(s)L(s) \quad \text{for} \quad s > 1,$$

where

$$L(s) = \sum_{m=0}^{\infty} \frac{(-1)^m}{(2m + 1)^s} \quad \text{for} \quad s > 1.$$

5.21. For the arithmetical functions defined in Exercises 1.92, 1.93 and 1.95,

$$\sum_{n=1}^{\infty} \frac{\lambda_{k,q}(n)}{n^s} = \frac{\zeta(ks)}{\zeta(qs)} \quad \text{for} \quad s > 1,$$

$$\sum_{n=1}^{\infty} \frac{\mu_{k,q}(n)}{n^s} = \frac{\zeta(qs)}{\zeta(ks)} \quad \text{for} \quad s > 1,$$

$$\sum_{n=1}^{\infty} \frac{\zeta_{k,q}(n)}{n^s} = \frac{\zeta(s)\zeta(ks)}{\zeta(qs)} \quad \text{for} \quad s > 1$$

and

$$\sum_{n=1}^{\infty} \frac{\phi_{k,q}(n)}{n^s} = \frac{\zeta(s - 1)\zeta(ks)}{\zeta(qs)} \quad \text{for} \quad s > 2.$$

(Hint: use Theorem 5.3 to obtain the Dirichlet series of $\lambda_{k,q}$.)

5.22. $$\sum_{n=1}^{\infty} \frac{\lambda(n)J_k(n)}{n^s} = \frac{\zeta(s)\zeta(2(s-k))}{\zeta(s-k)\zeta(2s)} \quad \text{for} \quad s > k+1.$$

5.23. $$\sum_{n=1}^{\infty} \frac{\lambda(n)\sigma_h(n)\sigma_k(n)}{n^s}$$

$$= \frac{\zeta(2s)\zeta(2(s-h))\zeta(2(s-k))\zeta(2(s-h-k))}{\zeta(s)\zeta(s-h)\zeta(s-k)\zeta(s-h-k)\zeta(2s-h-k)} \quad \text{for} \quad s > h+k+1.$$

5.24. Let k be a nonnegative integer and t a positive integer, and let

$$\rho'_{k,t}(n) = \sum_{\substack{d|n \\ d \text{ a } t\text{th power}}} d^k \quad \text{for all } n.$$

Then

$$\sum_{n=1}^{\infty} \frac{\lambda(n)\rho'_{k,t}(n)}{n^s} = \frac{\zeta(2s)\zeta(2t(s-k))}{\zeta(s)\zeta(t(s-k))} \quad \text{for} \quad s > k+1.$$

In particular,

$$\sum_{n=1}^{\infty} \frac{\lambda(n)\sigma_k(n)}{n^s} = \frac{\zeta(2s)\zeta(2(s-k))}{\zeta(s)\zeta(s-k)} \quad \text{for} \quad s > k+1.$$

5.25. Let $\bar{\mu}_k$ be the arithmetical function defined by

$$\bar{\mu}_k(n) = \begin{cases} \lambda(n) & \text{if } n \text{ is not divisible by the kth power of a prime} \\ 0 & \text{otherwise:} \end{cases}$$

then $\bar{\mu}_1 = \delta$ and $\bar{\mu}_2 = \mu$. For $s > 1$,

$$\sum_{n=1}^{\infty} \frac{\bar{\mu}_k(n)}{n^s} = \begin{cases} \dfrac{\zeta(2s)}{\zeta(s)\zeta(ks)} & \text{if } k \text{ is even} \\ \dfrac{\zeta(2s)\zeta(ks)}{\zeta(s)\zeta(2ks)} & \text{if } k \text{ is odd.} \end{cases}$$

5.26. $$\sum_{n=1}^{\infty} \frac{|\mu(n)|\phi(n)}{n^s} = \prod_p \frac{p^s + p - 1}{p^s} \quad \text{for } s > 1$$

and

$$\sum_{n=1}^{\infty} \frac{\gamma(n)}{n^s} = \zeta(s) \prod_p \frac{p^s + p - 1}{p^s} \quad \text{for } s > 1,$$

where γ is the core function defined in Exercise 1.14.

5.27. Let $\gamma'(n) = (-1)^{\omega(n)}\gamma(n)$. For all positive integers k,

$$\sum_{n=1}^{\infty} \frac{(-1)^{(k+1)\omega(n)}\gamma'(n)^k J_k(n)}{n^s} = \frac{\zeta(s-k)}{\zeta(s-2k)} \quad \text{for } s > 2k + 1.$$

In particular,

$$\sum_{n=1}^{\infty} \frac{\gamma'(n)\phi(n)}{n^s} = \frac{\zeta(s-1)}{\zeta(s-2)} \quad \text{for} \quad s > 3.$$

5.28. If k is an integer and $k \geq 2$ then

$$\sum_{n=1}^{\infty} \frac{\lambda(n)\tau_{k-1}(n)\tau(n^2\gamma(n)^{k-2})}{(k-1)^{\omega(n)} n^s} = \frac{\zeta(2s)^k}{\zeta(s)^{k+1}} \quad \text{for} \quad s > 1.$$

(τ_k is defined in Exercise 1.32 only for $k \geq 2$: let $\tau_1 = \zeta$.)

In particular,

$$\sum_{n=1}^{\infty} \frac{\lambda(n)\tau(n^2)}{n^s} = \frac{\zeta(2s)^2}{\zeta(s)^3} \quad \text{for} \quad s > 1$$

and

$$\sum_{n=1}^{\infty} \frac{\lambda(n)\tau(n)^2}{n^s} = \frac{\zeta(2s)^3}{\zeta(s)^4} \quad \text{for} \quad s > 1.$$

5.29. If k is an integer and $k \geq 3$ then

$$\sum_{n=1}^{\infty} \frac{\tau_{k-2}(n)\tau(n^2\gamma(n)^{k-3})}{(k-2)^{\omega(n)} n^s} = \frac{\zeta(s)^k}{\zeta(2s)} \quad \text{for} \quad s > 1.$$

In particular,

$$\sum_{n=1}^{\infty} \frac{\tau(n^2)}{n^s} = \frac{\zeta(s)^3}{\zeta(2s)} \quad \text{for} \quad s > 1$$

and

$$\sum_{n=1}^{\infty} \frac{\tau(n)^2}{n^s} = \frac{\zeta(s)^4}{\zeta(2s)} \quad \text{for} \quad s > 1$$

(see Exercise 5.33).

5.30. Let k be a nonnegative integer and let σ'_k be the arithmetical function defined by

$$\sigma'_k(n) = \sum_{d|n} \lambda(d)d^k \quad \text{for all} \quad n.$$

Then

$$\sum_{n=1}^{\infty} \frac{\sigma'_k(n)}{n^s} = \frac{\zeta(s)\zeta(2(s-k))}{\zeta(s-k)} \quad \text{for} \quad s > k+1.$$

5.31. For all n,

$$\sigma'_k(n) = \sum_{d|n} \nu_2(d)\lambda(n/d)J_k(n/d).$$

5.32. If k is a positive integer then

$$\sum_{n=1}^{\infty} \frac{\tau(n)\sigma_k(n)}{n^s} = \frac{\zeta(s)^2\zeta(s-k)^2}{\zeta(2s-k)} \quad \text{for} \quad s > k+1.$$

This can be proved in several ways, and in particular by using Theorem 5.3.

5.33. Theorem 5.3 can be used to prove that

$$\sum_{n=1}^{\infty} \frac{\tau(n)^2}{n^s} = \frac{\zeta(s)^4}{\zeta(2s)} \quad \text{for} \quad s > 1.$$

It can also be proved by using Theorem 5.2 (Hint: examine $\theta * \tau$), and it was part of Exercise 5.29.

5.34. $$\sum_{n=1}^{\infty} \frac{\beta_h(n)\sigma_k(n)}{n^s}$$

$$= \frac{\zeta(2s)\zeta(s-h)\zeta(s-h-k)\zeta(2(s-k))\zeta(2s-h-k)}{\zeta(s)\zeta(s-k)\zeta(2(2s-h-k))} \quad \text{for } s > h+k+1.$$

5.35. $$\sum_{n=1}^{\infty} \frac{\sigma_h'(n)\sigma_k(n)}{n^s}$$

$$= \frac{\zeta(s)\zeta(2(s-h))\zeta(s-k)\zeta(2(s-h-k))\zeta(2s-h-k)}{\zeta(s-h)\zeta(s-h-k)\zeta(2(2s-h-k))} \quad \text{for } s > h+k+1.$$

5.36. $$\sum_{n=1}^{\infty} \frac{\lambda(n)\sigma_h'(n)\sigma_k(n)}{n^s}$$

$$= \frac{\zeta(2s)\zeta(2(s - k))\zeta(s - h)\zeta(2s - h - k)}{\zeta(s)\zeta(2(2s - h - k))} \quad \text{for} \quad s > h + k + 1.$$

5.37. $$\sum_{n=1}^{\infty} \frac{\sigma_h'(n)\sigma_k'(n)}{n^s}$$

$$= \frac{\zeta(s)\zeta(2(s - h))\zeta(2(s - k))\zeta(s - h - k)}{\zeta(2s - h - k)\zeta(s - h)\zeta(s - k)} \quad \text{for} \quad s > h + k + 1.$$

5.38. $$\sum_{n=1}^{\infty} \frac{\lambda(n)\sigma_h'(n)\sigma_k'(n)}{n^s}$$

$$= \frac{\zeta(2s)\zeta(2(s - h - k))\zeta(s - h)\zeta(s - k)}{\zeta(s)\zeta(s - h - k)\zeta(2s - h - k)} \quad \text{for} \quad s > h + k + 1.$$

5.39. $$\sum_{n=1}^{\infty} \frac{\sigma_k(n^2)}{n^s} = \frac{\zeta(s)\zeta(s - k)\zeta(s - 2k)}{\zeta(2(s - k))}$$

and

$$\sum_{n=1}^{\infty} \frac{\sigma_k'(n^2)}{n^s} = \frac{\zeta(s)\zeta(s - 2k)}{\zeta(s - k)}$$

for $s > 2k + 1$. (Hint: let $h = 0$ in Exercises 5.35 and 5.37.)

5.40. For all n,

$$\sigma_k(n)^2 = \sum_{d|n} d^k \sigma_k(n^2/d^2)$$

and

$$\sigma_k'(n)^2 = \sum_{d|n} \lambda(d) d^k \sigma_k'(n^2/d^2).$$

These identities hold because σ_k and σ_k' are specially multiplicative functions (see Theorem 1.12). They can be proved using the results of the preceding exercise.

5.41. (A more general version of the result of Theorem 5.3.) Let f be a multiplicative function such that $\sum_{n=1}^{\infty} f(n)$ converges absolutely. Then, for each prime p, the series

$$1 + f(p) + f(p^2) + \ldots$$

converges absolutely and

$$\sum_{n=1}^{\infty} f(n) = \prod_p (1 + f(p) + f(p^2) + \ldots),$$

where the infinite product converges absolutely. If f is completely multiplicative then

$$\sum_{n=1}^{\infty} f(n) = \prod_{p} \frac{1}{1 - f(p)} .$$

5.42. Let f be a specially multiplicative function and $B = B_f$. Suppose that the Dirichlet series of f and B converge absolutely for $s > s_0$. Then, for $s > s_0$,

$$\sum_{n=1}^{\infty} \frac{\lambda(n)f(n)}{n^s} = \prod_{p} \frac{1}{1 + f(p)p^{-s} + B(p)p^{-2s}}$$

and

$$\sum_{n=1}^{\infty} \frac{N(f)(n)}{n^s} = \prod_{p} \frac{1}{1 - (f(p)^2 - 2B(p))p^{-s} + B(p)p^{-2s}}$$

($N(f)$ is defined in Exercise 1.70.)

5.43. Continuing Exercise 5.42, let F' and G be the generating functions of $N(f)$ and B, respectively. For $s > s_0$,

$$\sum_{n=1}^{\infty} \frac{f(n)^2}{n^s} = F'(s)G(s) \prod_{p} (1 + B(p)p^{-s})$$

and

$$\sum_{n=1}^{\infty} \frac{f(n^2)}{n^s} = F'(s) \prod_{p} (1 + B(p)p^{-s}).$$

5.44. Let f and g be arithmetical functions and let $\bar{f} = f * \zeta$ and $\bar{g} = g * \zeta$. Then

$$\sum_{n=1}^{\infty} \frac{\bar{f}(n)\bar{g}(n)}{n^s} = \zeta(s) \sum_{n=1}^{\infty} \frac{\phi_s(n)}{n^{2s}} \sum_{j=1}^{\infty} \frac{f(nj)}{j^s} \sum_{k=1}^{\infty} \frac{g(nk)}{k^s} \quad \text{for} \quad s > s_0,$$

if all of the series converge absolutely for $s > s_0$. The right-hand side is equal to $\zeta(s)H(s)$, where $H(s)$ is the generating function of the arithmetical function h in Exercise 1.4.

5.45. In the preceding exercise, if $f = \zeta_h$ and $h = \zeta_k$, the result is that of Proposition 5.4. (Hint: first show that if t is a real number then

$$\sum_{n=1}^{\infty} \frac{\phi_t(n)}{n^s} = \frac{\zeta(s - t)}{\zeta(s)} \quad \text{for} \quad s > t + 1:$$

see Exercise 5.6.)

5.46. The results in Exercises 5.35 and 5.37 are special cases of the formula in Exercise 5.44.

5.47. Assume that the Dirichlet series of the arithmetical function f converges absolutely for $s > s_0$. Then, for a positive integer r,

$$\sum_{n=1}^{\infty} \frac{f(n)c(n,r)}{n^s} = \sum_{d|r} \left(d^{1-s}\mu(r/d) \sum_{m=1}^{\infty} \frac{f(md)}{m^s} \right) \quad \text{for} \quad s > s_0.$$

In particular, if f is completely multiplicative then

$$\sum_{n=1}^{\infty} \frac{f(n)c(n,r)}{n^s} = \left(\sum_{m=1}^{\infty} \frac{f(m)}{m^s}\right)\left(\sum_{d|r} d^{1-s} f(d)\mu(r/d)\right) \quad \text{for} \quad s > s_0.$$

5.48. $$\sum_{n=1}^{\infty} \frac{\lambda(n)c(n,r)}{n^s} = \frac{\zeta(2s)}{\zeta(s)} \sum_{d|r} d^{1-s}\lambda(d)\mu(r/d) \quad \text{for} \quad s > 1.$$

5.49 $$\sum_{n=1}^{\infty} \frac{c(kn,r)}{n^s} = \zeta(s) \sum_{d|r} d^{1-s}(k,d)^s \mu(r/d) \quad \text{for } s > 1 .$$

5.50. Let $c_k(n,r)$ be the generalized Ramanujan sum defined in Exercise 2.51. For a real number s let $\sigma_s^{(k)}$ be the arithmetical function defined by

$$\sigma_s^{(k)}(n) = \sum_{d^k|n} d^{ks} \quad \text{for all } n.$$

Then, for $s > 1$,

$$\sigma_{1-s}^{(k)}(n) = \zeta(ks) \sum_{r=1}^{\infty} \frac{c_k(n,r)}{r^{ks}} \quad \text{for all } n .$$

5.51. For $s > 1$,

$$\phi_{k(1-s)}(r) = \frac{1}{\zeta(s)} \sum_{n=1}^{\infty} \frac{c_k(n,r)}{n^s} \quad \text{for all } r .$$

5.52. Let r be a positive integer. If f is an even function (mod r) with Fourier coefficients $\alpha(d)$, then

$$\sum_{n=1}^{\infty} \frac{f(n)}{n^s} = \zeta(s) \sum_{d|r} \alpha(d)\phi_{1-s}(d) \quad \text{for } s > 1 .$$

5.53. Let A be a regular arithmetical convolution. For a real number s let $\phi_{A,s}$ be the arithmetical function defined by

$$\phi_{A,s}(r) = \sum_{d \in A(r)} d^s \mu_A(r/d) .$$

Then

$$\sum_{n=1}^{\infty} \frac{c_A(n,r)}{n^s} = \zeta(s)\, \phi_{A,1-s}(r) \quad \text{for} \quad s > 1 .$$

5.54. $$\sum_{n=1}^{\infty} \frac{c(n,r)^2}{n^s} = \zeta(s) \sum_{d|r} \theta(r/d,r)\phi_{1-s}(d) \quad \text{for} \quad s > 1 .$$

(Hint: see Exercise 3.20.)

5.55. $$\sum_{n=1}^{\infty} \frac{S(n,r,t)}{n^s} = r^{2t-1}\zeta(s) \sum_{d|r} \frac{1}{d^t} \phi_{1-s}(d) \quad \text{for} \quad s > 1$$

(see Theorem 3.8).

5.56. Let r be a positive integer. If f is an even function (mod r) and if g is the function of two positive integer variables in Theorem 2.10, then

$$\sum_{n=1}^{\infty} \frac{f(n)}{n^s} = \zeta(s) \sum_{d|r} g(d,r/d)d^{-s} \quad \text{for} \quad s > 1 .$$

5.57. Let g and h be arithmetical functions and let the sum $f(n,r)$ be defined by

$$f(n,r) = \sum_{d|(n,r)} g(d)h(r/d) \quad \text{for all } n \text{ and } r .$$

Then, for all r ,

$$\sum_{n=1}^{\infty} \frac{f(n,r)}{n^s} = \zeta(s) \sum_{d|r} g(d)h(r/d)d^{-s} \quad \text{for } s > 1 .$$

If s is a real number such that $\sum_{r=1}^{\infty} h(r)/r^s$ converges, say to $H(s)$, then for all n ,

$$\sum_{r=1}^{\infty} \frac{f(n,r)}{r^s} = H(s) \sum_{d|n} \frac{g(d)}{d^s} .$$

5.58. $$\sum_{r=1}^{\infty} \frac{S(n,r,t)}{r^s} = \frac{\zeta(s-2t+1)\sigma_{s-t}(n)}{\zeta(s-t+1)n^{s-t}} \quad \text{for } s > 2t .$$

5.59. $$\sum_{r=1}^{\infty} \frac{D_{k,q}(n,r)}{r^s} = \frac{\sigma_{s-1}(n)\zeta(ks)}{n^{s-1}\ \zeta(qs)} \quad \text{for } s > 1 .$$

($D_{k,q}(n,r)$ is defined in Exercise 3.9.)

5.60 Let f and g be as in the first sentence of Theorem 5.7, and suppose that for an arithmetical function h ,

$$f(n) = \sum_{r=1}^{\infty} h(r)c(n,r) \quad \text{for all } n$$

and $\sum_{n=1}^{\infty} h(n)$ converges absolutely. Then

$$h(n) = \sum_{m=1}^{\infty} g(mn) \quad \text{for all } n.$$

5.61. Let f, g and h be arithmetical functions and assume that $h(1) \neq 0$ and that

$$\sum_{k=1}^{\infty} \sum_{m=1}^{\infty} h(k)h^{-1}(m)g(kmn)$$

converges absolutely for all n. If

$$f(n) = \sum_{m=1}^{\infty} h(m)g(mn) \quad \text{for all } n$$

then

$$g(n) = \sum_{m=1}^{\infty} h^{-1}(m)f(mn) \quad \text{for all } n.$$

5.62. For a real number s let β_s be the arithemtical function defined by

$$\beta_s(n) = \sum_{d|n} d^s \lambda(n/d) \quad \text{for all } n.$$

(If s is a positive integer this is the function defined in Exercise 1.78.) For $s > 0$,

$$\beta_s(n) = \frac{\zeta(2(s+1))}{\zeta(s+1)} \sum_{r=1}^{\infty} \frac{\lambda(r)}{r^{s+1}} c(n,r) \quad \text{for all } n.$$

In particular,

$$\beta(n) = \frac{\pi^2}{15} \sum_{r=1}^{\infty} \frac{\lambda(r)}{r^2} c(n,r) \quad \text{for all } n .$$

5.63. Let n be an integer. If

$$F(s) = \sum_{\substack{r=1 \\ (r,n)=1}}^{\infty} \frac{f(r)}{r^s} \quad \text{and} \quad G(s) = \sum_{\substack{r=1 \\ (r,n)=1}}^{\infty} \frac{g(r)}{r^s} ,$$

where the series converge absolutely for $s > s_0$, then

$$F(s)G(s) = \sum_{\substack{r=1 \\ (r,n)=1}}^{\infty} \frac{(f * g)(r)}{r^s} \quad \text{for } s > s_0 .$$

5.64. For $s > 1$,

$$\frac{n^s}{\psi_s(n)} = \frac{\zeta(2s)}{\zeta(s)} \sum_{\substack{r=1 \\ (r,n)=1}}^{\infty} \frac{|\mu(r)|}{r^s} \quad \text{for all } n .$$

5.65. For $s > 1$,

$$\frac{\phi_s(n)}{\psi_s(n)} = \frac{\zeta(2s)}{\zeta(s)^2} \sum_{\substack{r=1 \\ (r,n)=1}}^{\infty} \frac{\theta(r)}{r^s} \quad \text{for all } n .$$

5.66. Let F be the generating function of the arithmetical function f:

$$F(s) = \sum_{n=1}^{\infty} \frac{f(n)}{n^s} \quad \text{for } s > s_a.$$

Then F is differentiable for $s > s_a$ and

$$F'(s) = -\sum_{n=1}^{\infty} \frac{f(n)\log n}{n^s} \quad \text{for } s > s_a.$$

In particular,

$$\sum_{n=1}^{\infty} \frac{\log n}{n^s} = -\zeta'(s) \quad \text{for } s > 1.$$

5.67. Mangoldt's function Λ is defined by

$$\Lambda(n) = \begin{cases} \log p & \text{if } n = p^{\alpha}, \text{ where } p \text{ is a prime and } \alpha \geq 1 \\ 0 & \text{otherwise.} \end{cases}$$

Then $(\zeta * \Lambda)(n) = \log n$ for all n and

$$\sum_{n=1}^{\infty} \frac{\Lambda(n)}{n^s} = -\frac{\zeta'(s)}{\zeta(s)} \quad \text{for } s > 1.$$

5.68. If the Dirichlet series of the completely multiplicative function f converges absolutely to $F(s)$ for $s > s_0$ then

$$\sum_{n=1}^{\infty} \frac{f(n)\Lambda(n)}{n^s} = -\frac{F'(s)}{F(s)} \quad \text{for } s > s_0.$$

5.69. For all r ,

$$\sum_{n=1}^{\infty} \frac{c(n,r)\log n}{n^s} = -\phi_{1-s}(r)\left\{\zeta'(s)+\zeta(s)\Big(\log r + \alpha(1-s,r)\Big)\right\} \text{ for } s > 1 ,$$

where

$$\alpha(s,r) = \sum_{p|r} \frac{\log p}{p^s - 1} .$$

(Hint: show that $\frac{d}{ds}\phi_s(r) = \phi_s(r) \quad (\log r + \alpha(s,r))$.)

5.70. For all r ,

$$\sum_{\substack{n=1\\(n,r)=1}}^{\infty} \frac{\mu(n)\log n}{n^s} = \frac{r^s}{\zeta(s)\phi_s(r)}\left\{\frac{\zeta'(s)}{\zeta(s)} + \alpha(s,r)\right\} \quad \text{for } s > 1 .$$

5.71. For all n ,

$$\sum_{\substack{r=1\\(r,n)=1}}^{\infty} \frac{\lambda(r)\log r}{r^s} = \frac{\zeta(2s)\psi_s(n)}{n^s\zeta(s)}\left\{\frac{\zeta'(s)}{\zeta(s)} - \frac{2\zeta'(2s)}{\zeta(2s)} + \alpha(s,n)\right\} \quad \text{for } s > 1 .$$

5.72. For all n ,

$$\sum_{\substack{r=1\\(r,n)=1}}^{\infty} \frac{\theta(r)\log r}{r^s} = \frac{\zeta(s)^2\phi_s(n)}{\zeta(2s)\psi_s(n)}\left\{\frac{2\zeta'(2s)}{\zeta(2s)} - \frac{\zeta'(s)}{\zeta(s)} - \sum_{p|n} \frac{2p^s\log p}{p^{2s} - 1}\right\} \text{ for } s > 1 .$$

5.73. If δ_k is the arithmetical function defined in Exercise 1.20 then

$$\sum_{n=1}^{\infty} \frac{\delta_k(n)}{n^s} = \frac{k\zeta(s-1)\phi_{s-1}(k)}{\phi_s(k)} \quad \text{for } s > 2 .$$

5.74. $$\sum_{n=1}^{\infty} \frac{\delta_k(n)\log n}{n^s} = -\frac{k\zeta(s-1)\phi_{s-1}(k)}{\phi_s(k)}\left\{\frac{\zeta'(s-1)}{\zeta(s-1)} + \alpha(s-1,k) - \alpha(s,k)\right\}$$

for $s > 2$.

5.75. The identity of Exercise 2.34, which holds when $k = s-1$, s real, can be combined with Proposition 5.10 to obtain still another proof of Theorem 5.5.

5.76. For $s > 1$,

$$\sum_{d|n} \frac{|\mu(d)|d}{\phi_s(d)} = \zeta(s) \sum_{r=1}^{\infty} \frac{|\mu(r)|}{r^s} c(n,r) .$$

(See Exercise 2.45: the identity holds for real $s > 1$.)

5.77. For $s > 2$,

$$\sum_{d|n} \frac{\mu(d)\phi(d)d}{\phi_s(d)} = \frac{\zeta(s)}{\zeta(s-1)} \sum_{r=1}^{\infty} \frac{\mu(r)\phi(r)}{rJ_{s-1}(r)} c(n,r) .$$

(See Exercise 2.46: the identity holds for real $s > 2$.)

5.78. For $s > 1$

$$\sum_{d|n} \frac{|\mu(d)|\phi(d)}{\phi_s(d)} = \frac{\zeta(s)}{\zeta(s+1)} \sum_{r=1}^{\infty} \frac{|\mu(r)|\phi(r)}{\phi_{s-1}(r)} c(n,r).$$

(See Exercise 2.47: the identity holds for real $s > 1$.)

5.79. For $k \geq 0$ and $\varepsilon > 0$,

$$\lim_{n\to\infty} \frac{\sigma_k(n)}{n^{k+\varepsilon}} = 0.$$

If $\varepsilon = 0$ the limit exists but is not equal to zero.

5.80. If Φ_k is Klee's function (see Exercise 1.29) then for $\varepsilon > 0$,

$$\lim_{n\to\infty} \frac{n^{1-\varepsilon}}{\Phi_k(n)} = 0.$$

Notes on Chapter 5

The chapter begins with a number of classical results, including many examples of generating functions in the text, in early exercises and in Exercises 5.66-5.68. There are other expositions of these results, and much more, in books by E. Landau [L] and T. M. Apostol [A]. In these books the variable s is, for the most part, a complex variable, and the techniques of complex function theory are used to study the Dirichlet series and the functions they represent.

Proposition 5.4 was stated by S. Ramanujan [16a], and his statement included the cases in which h and/or k equals zero, i.e., Exercises 5.32 and 5.33. The first published proof was by B. M. Wilson [23]: it is the proof given the text. The related results in Exercises 5.35-5.38, involving the function σ_k' in Exercise 5.30, were obtained by S. Chowla [28]. The formula in Exercise 5.44 and its application to the proof of Proposition 5.4, are due to D. M. Kotelyanskiǐ [53]. The proof of that same proposition using the identity in Exercise 1.63 was given by J. Lambek [66]. The generating functions of the products of three divisor sum functions and of two unitary divisor sum functions were found by M. V. Subbarao [68a]. An analogue of Proposition 5.4 involving basic sequences is in a paper by D. L. Goldsmith [69]: he used the identity in Exercise 4.47.

The generating function of Klee's function Φ_k in Exercise 5.12 was derived by U. V. Satyanarayana and K. Pattabhiramasastry [65], and those for the generalized Dedekind functions in Exercises 5.18 and 5.19 by D. Suryanarayana [69a] and J. Hanumanthachari [72], respectively. For the functions in Exercise 5.21 associated with the (k,q)-integers, the generating functions were given by M. V. Subbarao and V. C. Harris [66]. The generating function of the core function γ in Exercise 5.26 is in a paper by S. Wigert [32], and that of the function δ_k in Exercise 5.73 in one by D. Suryanarayana [69b]. The results in Exercises 5.42 and 5.43 involving the genrating functions associated with a specially multiplicative function are due to D. Redmond and R. Sivaramakrishnan [81], and other results along the same line are in a paper by A. Mercier [82]. The formulas in Exercises 5.27-5.29 were discovered by L. Gegenbauer, and can be found in the papers by him referred to by L. E. Dickson [D], Chapter X.

Theorems 5.5 and 5.8 were proved by S. Ramanujan [18], and he stated the formulas which express $\sigma(n)$ and $\phi(n)$ as series involving Ramanujan sums. He drew the conclusion that $\sum_{r=1}^{\infty} c(n,r)/r = 0$. The proof in the text, which assumes the deep fact that $\sum_{r=1}^{\infty} \mu(r)/r = 0$, is based on a theorem from §185 of the book by E. Landau [L].

S. Ramanujan [18] also gave the generating function of the arithmetical function $c(\cdot,r)$, in a note at the very end of his paper. The generalization to $c_A(\cdot,r)$ in Exercise 5.53 is new, as are the results in Exercises 5.52, 5.54-5.56 and 5.58. The first result in Exercise 5.57, which is a special case of the formula in Exercise 5.56, is due to D. R. Anderson and T. M. Apostol [53], and the second result of that exercise is in a paper by T. M. Apostol [72].

M. M. Crum [40] gave the results related to those of S. Ramanujan which are stated in Exercises 5.47-5.49, and the k-analogues of S. Ramanujan's formulas in Exercises 5.50 and 5.51 are due to E. Cohen [49], [56a]. The generating function in Exercise 5.59 was found by M. V. Subbarao and V. C. Harris [66]. Proposition 5.9, the examples which follow its proof and the examples in Exercises 5.64 and 5.65 are due to E. Cohen [61d].

The principle set down in Proposition 5.10 was done so by E. Cohen [59a]. He used it to give the proofs of Theorems 5.5 and 5.8

which follow the proof of the proposition, and in another paper [59d] he pointed out that it can be applied to derive the formulas in Exercises 5.76-5.78. Lemma 5.11 is Theorem 316 in the book by G. H. Hardy and E. M. Wright [HW].

In 1832, in a paper in volume 9 of Crelle's Journal, A. F. Möbius was led to the arithmetical function which bears his name by considering the following problem: if F and G are functions and if

$$F(x) = \sum_{m=1}^{\infty} a_m G(x^m),$$

determine numbers b_m such that

$$G(x) = \sum_{m=1}^{\infty} b_m F(x^m).$$

He argued formally, i.e., without regard to convergence, and showed that the numbers b_m are those determined by

$$\sum_{d|m} a_d b_{m/d} = \begin{cases} 1 & \text{if} \quad m = 1 \\ 0 & \text{if} \quad m \neq 1 \end{cases}$$

for all m. Thus, if $a_1 \neq 0$, the numbers b_m are the values of the arithmetical function inverse to the arithmetical function with values a_m.

If we define functions f and g by $f(x) = F(e^x)$ and $g(x) = G(e^x)$, and if we let x take on integer values, then A. F. Möbius' result becomes the following: if h is an arithmetical function with $h(1) \neq 0$, then

$$f(n) = \sum_{m=1}^{\infty} h(m)g(mn) \quad \text{for all} \quad n$$

if and only if

$$g(n) = \sum_{m=1}^{\infty} h^{-1}(m)f(mn) \quad \text{for all} \quad n.$$

Of course, this statement is not always true, even in the simplest case when $h = \zeta$, as we have seen in the remark following the proof of Proposition 5.6. The statement in Exercise 5.61, which has Proposition 5.6 as a corollary, was proved by E. Hille [37], and the aforementioned example was given by E. Hille and O. Szász [36a].

J. H. Laxton and J. W. Sanders [80] have written a short history of the Möbius inversion principle for infinite sums. They discussed some applications to numerical integration. The inversion principle has been applied to other problems in analysis by E. Hille and O. Szász [36a], [36b], O. Szász [47] and R. R. Goldberg and R. S. Varga [56]. The application of the inversion principle made in the text, namely Theorem 5.7, is due to D. Rearick [66a].

Chapter 6
Asymptotic Properties of Arithmetical Functions

Let us begin with an example. The object is to describe in some meaningful way the behavior of

$$\sum_{n \le x} \frac{1}{n}$$

as a function of the real variable x, for large x. To do this we need the following information.

Proposition 6.1. If, for each positive integer n,

$$c_n = 1 + \frac{1}{2} + \ldots + \frac{1}{n} - \log n ,$$

then the sequence $\{c_n\}$ converges.

Proof. For $k \ge 2$ let

$$A_k = \int_{k-1}^{k} \frac{1}{t}\, dt - \frac{1}{k} = \log k - \log (k-1) - \frac{1}{k} .$$

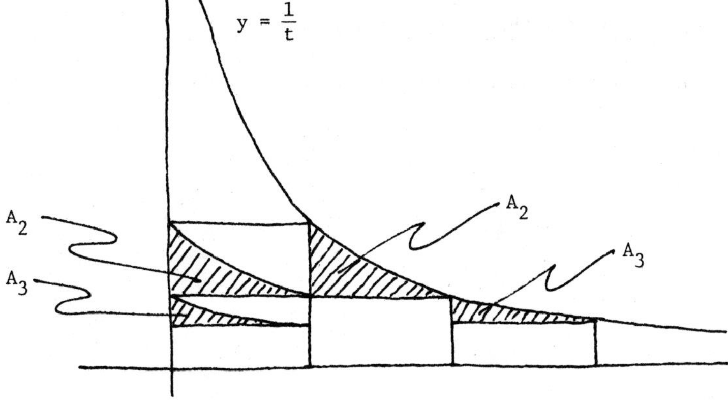

Then

$$\sum_{k=2}^{n} A_k = \log n - (1 + \frac{1}{2} + \ldots + \frac{1}{n}) + 1 = 1 - C_n .$$

In a square with unit sides there are nonoverlapping regions with areas $A_2, A_3, \ldots$. Thus $0 \leq 1 - C_n \leq 1$ for all n . Certainly $1 - C_n < 1 - C_{n+1}$ for all n . Hence, the sequence $\{1 - C_n\}$ is increasing and bounded, and consequently converges to a limit $1 - C$. Then

$$C = \lim_{n\to\infty} C_n . \square$$

The real number C is called <u>Euler's constant</u>. It is approximately 0.57721.

Note that

$$1 - C = \sum_{k=1}^{\infty} A_k$$

and that for all n ,

$$C_n - C = (1 - C) - (1 - C_n) = \sum_{k=n+1}^{\infty} A_k \leq \frac{1}{n}$$

(see the figure below for the case $n = 4$).

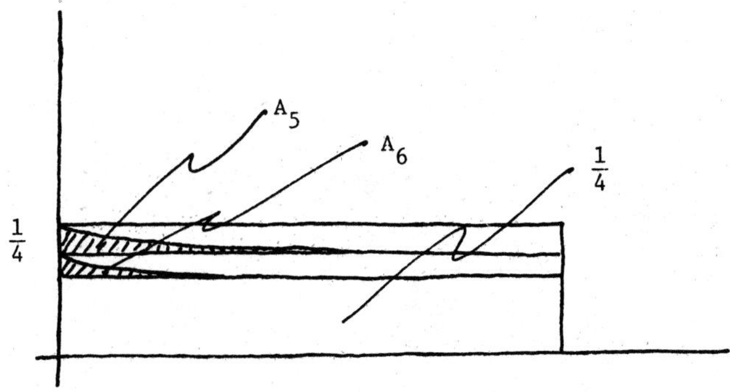

If $x \geq 1$ then

$$\sum_{n\leq x} \frac{1}{n} = \sum_{n=1}^{[x]} \frac{1}{n} = C_{[x]} + \log [x] .$$

Thus

$$\sum_{n\leq x} \frac{1}{n} - \log x - C = C_{[x]} - C + \log [x] - \log x$$

$$\leq \frac{1}{[x]} + \log \frac{[x]}{x} \leq \frac{x}{[x]} \cdot \frac{1}{x} \leq 2 \cdot \frac{1}{x} .$$

Therefore, the quotient

$$\frac{\displaystyle\sum_{n\leq x} \frac{1}{n} - \log x - C}{\frac{1}{x}}$$

is bounded for $x \geq 1$. This can be expressed conveniently by using the "big oh" notation of E. Landau.

Let $f(x)$ and $g(x)$ be real-valued functions of the real variable x, defined for all large x. Assume that $g(x)$ is positive for all large x. We write

$$f(x) = O(g(x)) ,$$

which is read " $f(x)$ is big oh of $g(x)$," if there is a real number K, independent of x, such that

$$\frac{|f(x)|}{g(x)} \leq K \qquad \text{for all large } x .$$

The expression $f_1(x) = f_2(x) + O(g(x))$ will mean that $f_1(x) - f_2(x) = O(g(x))$. In these terms we have

Proposition 6.2.

$$\sum_{n \leq x} \frac{1}{n} = \log x + C + O\left(\frac{1}{x}\right) .$$

There are results analogous to this one for powers of n other than n^{-1}. In deriving these results we will use the following fact.

The Euler Summation Formula. Let f be a real-valued function of the real variable x, and assume that f is defined and has a continuous derivative for $x > 0$. Then, for all $x \geq 1$,

$$\sum_{n \leq x} f(n) = f(1) + \int_1^x f(t)dt$$

$$+ \int_1^x (t-[t])f'(t)dt + ([x]-x)f(x) .$$

Proof. If $2 \le n \le x$ then

$$\int_{n-1}^n [t]\ f'(t)dt = (n-1) \int_{n-1}^n f'(t)dt$$

$$= nf(n) - (n-1)f(n-1) - f(n) .$$

Thus, writing m for $[x]$,

$$\sum_{n\le x} f(n) = f(1) + \sum_{n=2}^m f(n)$$

$$= f(1) + \sum_{n=2}^m (nf(n) - (n-1)f(n-1))$$

$$- \int_1^m [t]f'(t)dt$$

$$= mf(m) - \int_1^m [t]f'(t)dt .$$

Now,

$$\int_1^x (t-[t])f'(t)dt$$

$$= \int_1^x tf'(t)dt - \int_1^m [t]f'(t)dt - m\int_m^x f'(t)dt ,$$

and if we integrate the first integral by parts, this becomes

$$xf(x) - f(1) - \int_1^x f(t)dt - \int_1^m [t]f'(t)\,dt - mf(x) + mf(m) .$$

Therefore,

$$mf(m) - \int_1^m [t]f'(t)dt$$

$$= f(1) + \int_1^x f(t)dt + \int_1^x (t-[t])f'(t)dt$$

$$+ ([x]-x)f(x) . \square$$

Proposition 6.3. If $s > 1$ then

$$\sum_{n\leq x} \frac{1}{n^s} = \frac{x^{1-s}}{1-s} + \zeta(s) + O\left(\frac{1}{x^s}\right) .$$

Proof. Let $f(x) = \frac{1}{x^s}$. By the Euler summation formula

$$\sum_{n \le x} \frac{1}{n^s} = 1 + \int_1^x \frac{1}{t^s}\,dt - s\int_1^x \frac{t-[t]}{t^{s+1}}\,dt + \frac{[x]-x}{x^s}$$

$$= 1 + \frac{x^{1-s}}{1-s} - \frac{1}{1-s} - s\int_1^\infty \frac{t-[t]}{t^{s+1}}\,dt$$

$$+ s\int_x^\infty \frac{t-[t]}{t^{s+1}}\,dt + \frac{[x]-x}{x^s}\ .$$

Note that

$$\frac{[x]-x}{x^s} < \frac{1}{x^s} \quad \text{for all } x \ge 1\ .$$

Thus, by Exercise 6.5,

$$\sum_{n \le x} \frac{1}{n^s} = \frac{x^{1-s}}{1-s} + g(s) + 0\left(\frac{1}{x^s}\right),$$

where

$$g(s) = 1 - \frac{1}{1-s} - s\int_1^\infty \frac{t-[t]}{t^{s+1}}\,dt\ .$$

Now

$$\lim_{x\to\infty} \sum_{n\le x} \frac{1}{n^s} = \zeta(s) \quad \text{and} \quad \lim_{x\to\infty} \frac{x^{1-s}}{1-s} = 0 ,$$

and if $h(x) = O\left(\frac{1}{x^s}\right)$ then $\lim_{x\to\infty} h(x) = 0$. Therefore, $g(s) = \zeta(s)$. □

To each arithmetical function f there corresponds the <u>summatory function</u>

$$\sum_{n\le x} f(n)$$

of the real variable x . In this chapter we shall obtain formulas of the type given in Propositions 6.2 and 6.3 for the summatory functions of some of the arithmetical functions defined in earlier chapters. Formulas of this kind are called <u>asymptotic formulas</u>. They give some information about the behavior of the arithmetical functions as their arguments increase.

We have found asymptotic formulas for the summatory function of the function f defined by $f(n) = \frac{1}{n^s}$ when $s = 1$ and when $s > 1$. The case in which $s \le 0$ is treated in Exercise 6.9, and an asymptotic formula for the summatory function of the function g defined by $g(n) = \mu(n)/n^s$, for $s > 1$, is given in Exercise 6.10.

We are prepared now to prove the first of the principal results of the chapter.

<u>Theorem 6.4</u>. For a positive integer k ,

$$\sum_{n\le x} J_k(n) = \frac{x^{k+1}}{(k+1)\zeta(k+1)} + \begin{cases} O(x \log x) & \text{if } k = 1 \\ O(x^k) & \text{if } k \ge 2 . \end{cases}$$

Proof. Since $J_k = \mu * \zeta_k$ we have, by Exercise 6.14,

$$\sum_{n \le x} J_k(n) = \sum_{d \le x} \mu(d) \sum_{n \le \frac{x}{d}} n^k$$

$$= \sum_{d \le x} \mu(d) \left\{ \frac{1}{k+1} \left(\frac{x}{d}\right)^{k+1} + O\left(\left(\frac{x}{d}\right)^k\right) \right\}$$

by Exercise 6.9. Thus,

$$\sum_{n \le x} J_k(n) = \frac{x^{k+1}}{k+1} \sum_{d \le x} \frac{\mu(d)}{d^{k+1}} + \sum_{d \le x} O\left(\left(\frac{x}{d}\right)^k\right)$$

$$= \frac{x^{k+1}}{k+1} \sum_{d \le x} \frac{\mu(d)}{d^{k+1}} + O\left(\sum_{d \le x} \left(\frac{x}{d}\right)^k\right).$$

The last step, in which the summation sign and the big oh are interchanged, requires some explanation. By Exercise 6.9,

$$\sum_{n \le x} n^k = \frac{x^{k+1}}{k+1} + g(x)$$

where for some constant K,

$$|g(x)| \le Kx^k \quad \text{for all large } x .$$

Thus,

$$\sum_{d\leq x} O\left(\left(\frac{x}{d}\right)^k\right) = \sum_{d\leq x} g_d(x) ,$$

where $g_d(x) = g(x/d)$. Since $\frac{K}{d^k} \leq K$ for all $d \leq x$,

$$|g_d(x)| \leq Kx^k \quad \text{for all large } x \text{ , and all } d .$$

Therefore, by Exercise 6.4,

$$\sum_{d\leq x} O\left(\left(\frac{x}{d}\right)^k\right) = O\left(x^k \sum_{d\leq x} \frac{1}{d^k}\right).$$ [1)]

We must consider two cases. If $k = 1$ then by Proposition 6.2,

$$O\left(x \sum_{d\leq x} \frac{1}{d}\right) = O(x \log x + Cx + O(1)) = O(x \log x) .$$

If $k \geq 2$ then by Proposition 6.3,

$$O\left(x^k \sum_{d\leq x} \frac{1}{d^k}\right) = O\left(\frac{x}{1-k} + x^k\zeta(k) + O(1)\right) = O(x^k) .$$

Therefore,

$$\sum_{n\leq x} J_k(n) = \frac{x^{k+1}}{k+1} \sum_{d\leq x} \frac{\mu(d)}{d^{k+1}} + R(x) ,$$

[1)] A similar situation, in which we wish to interchange a summation sign and a big oh, will occur in other proofs. In each case the interchange can be justified in the same way as here, and the details will be left to the reader.

where

$$R(x) = \begin{cases} O(x \log x) & \text{if } k = 1 \\ O(x^k) & \text{if } k \geq 2 , \end{cases}$$

and so

$$\sum_{n \leq x} J_k(n) = \frac{x^{k+1}}{k+1}\left(\sum_{d=1}^{\infty} \frac{\mu(d)}{d^{k+1}} - \sum_{d>x} \frac{\mu(d)}{d^{k+1}}\right) + R(x)$$

$$= \frac{x^{k+1}}{(k+1)\zeta(k+1)} + \frac{x^{k+1}}{k+1} O(x^{-k}) + R(x) ,$$

by Exercise 6.10. The middle term is $O(x)$. Thus,

$$\sum_{n \leq x} J_k(n) = \frac{x^{k+1}}{(k+1)\zeta(k+1)} + \begin{cases} O(x \log x) & \text{if } k = 1 \\ O(x^k) & \text{if } k \geq 2 . \end{cases} \square$$

In particular, since $\zeta(2) = \pi^2/6$,

$$\sum_{n \leq x} \phi(n) = \frac{3x^2}{\pi^2} + O(x \log x) .$$

Almost the same proof can be used for asymptotic formulas for the summatory functions of the arithmetical functions σ_k, $k=1,2,\ldots$.

Theorem 6.5. For a positive integer k ,

$$\sum_{n\le x} \sigma_k(n) = \frac{\zeta(k+1)}{k+1} x^{k+1} + \begin{cases} O(x \log x) & \text{if } k = 1 \\ O(x^k) & \text{if } k \ge 2 . \end{cases}$$

Proof By Exercise 6.14,

$$\sum_{n\le x} \sigma_k(n) = \sum_{d\le x} \sum_{n\le \frac{x}{d}} n^k$$

$$= \sum_{d\le x} \left\{ \frac{1}{k+1} \left(\frac{x}{d}\right)^{k+1} + O\left(\left(\frac{x}{d}\right)^k\right) \right\}$$

$$= \frac{x^{k+1}}{k+1} \sum_{d\le x} \frac{1}{d^{k+1}} + O\left(x^k \sum_{d\le x} \frac{1}{d^k}\right) ,$$

where we have used Exercise 6.9. Thus, just as in the preceding proof,

$$\sum_{n\le x} \sigma_k(n) = \frac{x^{k+1}}{k+1} \sum_{d\le x} \frac{1}{d^{k+1}} + R(x) ,$$

where

$$R(x) = \begin{cases} O(x \log x) & \text{if } k = 1 \\ O(x^k) & \text{if } k \ge 2 . \end{cases}$$

Therefore,

$$\sum_{n\leq x} \sigma_k(n) = \frac{x^{k+1}}{k+1}\zeta(k+1) + \frac{x^{k+1}}{k+1}\sum_{d>x}\frac{1}{d^{k+1}} + R(x)$$

By Exercise 6.8, the middle term is

$$\frac{x^{k+1}}{k+1}\,O(x^{-k}) = O(x) ,$$

and the formula of the theorem follows. □

In particular,

$$\sum_{n\leq x} \sigma(x) = \frac{\pi^2 x^2}{12} + O(x \log x) .$$

The same kind of proof can be used to show that

$$\sum_{n\leq x} \tau(n) = x \log x + O(x) .$$

In fact, since $\tau = \zeta * \zeta$,

$$\sum_{n\leq x} \tau(n) = \sum_{d\leq x}\ \sum_{n\leq \frac{x}{d}} 1$$

$$= \sum_{d\leq x}\left(\frac{x}{d} + O(1)\right) = x\sum_{d\leq x}\frac{1}{d} + O(x) .$$

By Proposition 6.2 this is equal to

$$x \log x + Cx + O(1) + O(x)$$

$$= x \log x + O(x) .$$

A very different proof yields a stronger result, stronger in the sense that the error term $O(x)$ is replaced by $O(\sqrt{x})$.

Theorem 6.6.

$$\sum_{n \leq x} \tau(n) = x \log x + (2C - 1)x + O(\sqrt{x})$$

Proof. By Exercise 1.2,

$$\sum_{n \leq x} \tau(n) = \sum_{j \leq x} \left[\frac{x}{j}\right] .$$ [1)]

For a fixed x let T be that part of the st-plane bounded by the nonegative coordinate axes and the graph of $st = x$, $t > 0$. Then $\left[\frac{x}{j}\right]$ is the number of points

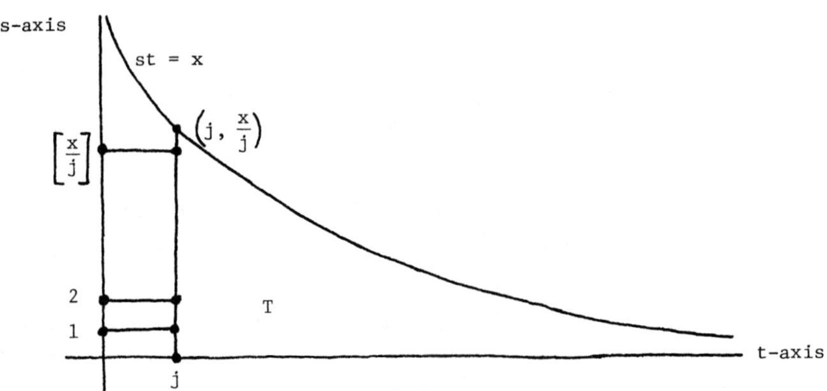

1) According to Exercise 1.2, $\sum_{n \leq x} \tau(n) = \sum_{j=1}^{[x]} \left[\frac{[x]}{j}\right]$. However, $\left[\frac{[x]}{j}\right] = \left[\frac{x}{j}\right]$ by Exercise 6.1(d).

(j,s) in T with s a positive integer. Thus

$$\sum_{n\leq x} \tau(n) = \text{the number of points in } T \text{ having both coordinates positive integers.}$$

We can count these points in another way. Their number is twice the number of such points in T above the line $s = t$

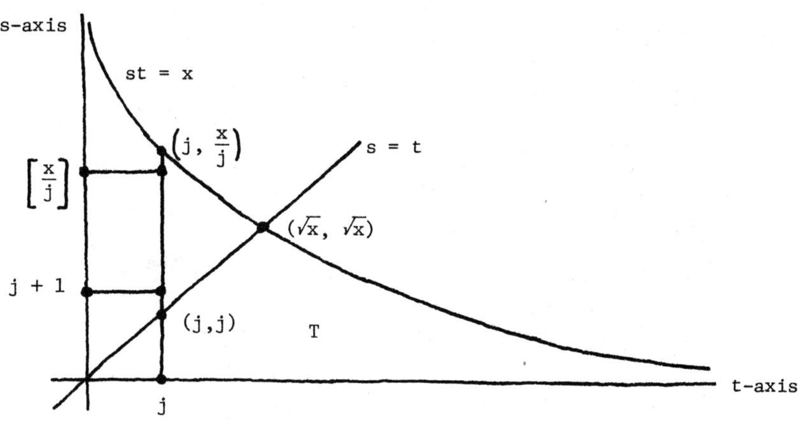

plus the number of such points on the line. Thus,

$$\sum_{n\leq x} \tau(n) = 2 \sum_{j\leq\sqrt{x}} \left(\left[\frac{x}{j}\right] - j\right) + [\sqrt{x}]$$

$$= 2 \sum_{j\leq\sqrt{x}} \left(\frac{x}{j} + O(1)\right) - [\sqrt{x}]([\sqrt{x}] + 1) + [\sqrt{x}]$$

$$= 2x \sum_{j \leq \sqrt{x}} \frac{1}{j} + O(\sqrt{x}) + [\sqrt{x}]^2 .$$

Since $[\sqrt{x}] = x + O(\sqrt{x})$ we have, by Proposition 6.2,

$$\sum_{n \leq x} \tau(n) = 2x\left(\log \sqrt{x} + C + O\left(\frac{1}{\sqrt{x}}\right)\right) - x + O(\sqrt{x})$$

$$= x \log x + (2C - 1)x + O(\sqrt{x}) . \square$$

The values of an arithmetical function can fluctuate wildly as its argument n increases. Consider for example the function τ. For every positive integer n_0 there are integers $m, n > n_0$ such that $\tau(m)$ is as small as possible and $\tau(n)$ is greater than any prescribed integer N. For if p is a prime with $p > n_0$ and $\alpha + 1 > N$ then

$$\tau(p) = 2 \quad \text{and} \quad \tau(p^\alpha) = \alpha + 1 > N .$$

Thus, rather than considering the values $\tau(n)$ themselves, it is perhaps better to investigate the average value

$$\frac{1}{x} \sum_{n \leq x} \tau(n)$$

of the function τ for integers $n \leq x$. Since

$$\sum_{n \le x} \tau(n) = x \log x + O(x) ,$$

this is equal to $\log x + g(x)$, where $g(x)$ is a bounded function, i.e.,

$$\frac{1}{x} \sum_{n \le x} \tau(n) - \log x$$

is bounded. Therefore, we can draw the weaker conclusion that

$$\lim_{x \to \infty} \frac{\frac{1}{x} \sum_{n \le x} \tau(n)}{\log x} = 1 .$$

If f and g are real-valued functions of a real variable x , defined for all large x , and if

$$\lim_{x \to \infty} \frac{f(x)}{g(x)} = 1 ,$$

we write $f(x) \sim g(x)$ and say that $f(x)$ is <u>asymptotic to</u> $g(x)$ (as $x \to \infty$) . This is another notational convention introduced by E. Landau. We have shown that

$$\frac{1}{x} \sum_{n \le x} \tau(n) \sim \log x ,$$

and we see from Theorem 6.4 and 6.5 that

$$\frac{1}{x} \sum_{n \le x} \phi(n) \sim \frac{3x}{\pi^2}$$

and

$$\frac{1}{x} \sum_{n \le x} \sigma(n) \sim \frac{\pi^2 x}{12} .$$

If k is any positive integer then

$$\frac{1}{x} \sum_{n \le x} J_k(n) \sim \frac{x^k}{(k+1)\zeta(k+1)}$$

and

$$\frac{1}{x} \sum_{n \le x} \sigma_k(n) \sim \frac{\zeta(k+1)x^k}{k+1} .$$

For certain arithmetical functions f,

$$\lim_{x \to \infty} \frac{1}{x} \sum_{n \le x} f(n)$$

exists. It is called the <u>mean value</u> of f and will be denoted by $M(f)$. For example, let $f(n) = \mu(n)/n^2$ for all n. By Exercise 6.10,

$$\sum_{n\leq x} \frac{\mu(n)}{n^2} = \frac{6}{\pi^2} + O\left(\frac{1}{x}\right).$$

Therefore, $M(f) = 0$. (See Exercise 6.10.)

The problem of determining which arithmetical funcions have a mean value is a difficult one, and is not treated in this book. We mention in passing that $M(\mu) = 0$.[1)]

Let U be the unitary convolution, defined in Chapter 4. Our goal now is to find asymptotic formulas for

$$\sum_{n\leq x} \phi_U(n) \quad \text{and} \quad \sum_{n\leq x} \sigma_U(n) ,$$

where ϕ_U and σ_U are defined by

$$\phi_U(n) = \sum_{d\in U(n)} d\mu_U(n/d) = \sum_{\substack{d|n \\ (d,n/d)=1}} d\mu_U(n/d)$$

and

$$\sigma_U(n) = \sum_{d\in U(n)} d = \sum_{\substack{d|n \\ (d,n/d)=1}} d .$$

---

1) This fact is equivalent to the prime number theorem, and consequently it is equivalent to the fact that $\sum_{n=1}^{\infty} \mu(n)/n = 0$. See [Ay] Chap. II, especially pp. 113-116.

Both of these functions fit into the general framework of a function f defined by

$$f(n) = \sum_{\substack{d|n \\ (d,n/d)=1}} d\, g(n/d) ,$$

where g is an arithmetical function.

Theorem 6.7. If the function g is bounded then

$$\sum_{n\leq x} f(n) = \frac{x^2}{2} \sum_{n=1}^{\infty} \frac{g(n)\phi(n)}{n^3} + O(x(\log x)^2) .$$

If g is bounded then the series

$$\sum_{n=1}^{\infty} \frac{g(n)\phi(n)}{n^3}$$

converges absolutely because the Dirichlet series of ϕ converges absolutely for $s > 2$. If $g = \zeta$, it converges to $\zeta(2)/\zeta(3) = \pi^2/6\zeta(3)$. If $g = \mu_U$, it converges to a real number

$$A = \sum_{n=1}^{\infty} \frac{\mu_U(n)\phi(n)}{n^3} .$$

If p is a prime then

$$\sum_{j=0}^{\infty} \frac{\mu_U(p^j)\phi(p^j)}{p^{3j}} = 1 - \sum_{j=1}^{\infty} \frac{\phi(p^j)}{p^{3j}} = 1 - \sum_{j=1}^{\infty} \frac{p^j(1-\frac{1}{p})}{p^{3j}}$$

$$= 1 - \frac{p-1}{p} \sum_{j=1}^{\infty} \left(\frac{1}{p^2}\right)^j = 1 - \frac{1}{p^2+p} .$$

Therefore, by Theorem 5.6,

$$A = \prod_p \left(1 - \frac{1}{p^2+p}\right)$$

(see also Exercise 6.23).

<u>Corollary 6.8.</u> $\sum_{n \leq x} \phi_U(n) = \frac{Ax^2}{2} + O(x(\log x)^2)$.

<u>Corollary 6.9.</u> $\sum_{n \leq x} \sigma_U(n) = \frac{\pi^2 x^2}{12\zeta(3)} + O(x(\log x)^2)$.

In the course of proving Theorem 6.7 we will need to know that if n is a positive integer then

$$\sum_{\substack{e \leq x \\ (e,n)=1}} e = \frac{\phi(n)x^2}{2n} + O(x\theta(n)).$$ [1)]

By Exercise 6.24,

---

1) $f(x) = O(g(x,n))$ means that $|f(x)|/g(x,n)$ is bounded uniformly in x for all n , i.e., the bounding constant is independent of both x and n .

$$\sum_{\substack{e \le x \\ (e,n)=1}} e = \sum_{e \le x} e\,\delta((e,n))$$

$$= -\sum_{e \le x} \sum_{k \le e} \delta((k,n)) + ([x] + 1) \sum_{e \le x} \delta((e,n))$$

$$= -\sum_{e \le x} \phi(e,n) + ([x] + 1)\phi(x,n) \ .$$

The function $\phi(x,n)$ was defined in Exercise 1.28, and by that exercise,

$$\phi(x,n) = \frac{\phi(n)x}{n} + O(\theta(n)) \ .$$

Thus,

$$([x] + 1)\phi(x,n) = \frac{\phi(n)x^2}{n} + \frac{\phi(n)x}{n} + O(x\theta(n)) + O(\theta(n))$$

$$= \frac{\phi(n)x^2}{n} + O(x\theta(n)) \ .$$

Also,

$$-\sum_{e \le x} \phi(e,n) = -\sum_{e \le x} \left(\frac{\phi(n)e}{n} + O(\theta(n)) \right)$$

$$= -\frac{\phi(n)}{n} \sum_{e \le x} e + O(x\theta(n))$$

$$= -\frac{\phi(n)}{n}\left(\frac{x^2}{2} + O(x)\right) + O(x\theta(n))$$

by Exercise 6.9. Since $\phi(n)/n \leq 1 \leq \theta(n)$ for all n, this is equal to

$$-\frac{\phi(n)x^2}{2n} + O(x\theta(n)) ,$$

and the required result follows.

Proof of Theorem 6.7.

$$\sum_{n\leq x} f(n) = \sum_{n\leq x} \sum_{\substack{d|n\\(d,n/d)=1}} g(d)\frac{n}{d}$$

$$= \sum_{d\leq x} g(d) \sum_{\substack{e\leq\frac{x}{d}\\(e,d)=1}} e$$

$$= \sum_{d\leq x} g(d)\left\{\frac{\phi(d)x^2}{2d^3} + O\left(\frac{x}{d}\,\theta(d)\right)\right\}$$

$$= \frac{x^2}{2}\sum_{d\leq x}\frac{g(d)\phi(d)}{d^3} + O\left(x\sum_{d\leq x}\frac{\theta(d)}{d}\right) .$$

Now

$$\sum_{d\leq x}\frac{\theta(d)}{d} \leq \sum_{d\leq x}\frac{\tau(d)}{d} = O((\log x)^2)$$

by Exercise 6.16. Thus

$$\sum_{n\leq x} f(n) = \frac{x^2}{2}\sum_{n=1}^{\infty}\frac{g(n)\phi(n)}{n^3} - \frac{x^2}{2}\sum_{n>x}\frac{g(n)\phi(n)}{n^3} + O(x(\log x)^2) .$$

The second sum is

$$O\left(x^2 \sum_{n>x}\frac{\phi(n)}{n^3}\right) ,$$

and since $\phi(n) \leq n$ for all n ,

$$\sum_{n>x}\frac{\phi(n)}{n^3} = O\left(\sum_{n>x}\frac{1}{n^2}\right) = O\left(\frac{1}{x}\right)$$

by Exercise 6.8. Therefore,

$$\frac{x^2}{2}\sum_{n>x}\frac{g(n)\phi(n)}{n^3} = O(x) = O(x(\log x)^2) ,$$

and the proof of the theorem is complete. □

As another example of a general result that has special cases of interest, we shall obtain an asymptotic formula for the summatory function of an arithmetical function f given by

$$f(n) = \sum_{\substack{r=1\\(r,n)=1}}^{\infty}\frac{g(r)}{r^s} \qquad \text{for} \quad s > 1 ,$$

for some arithmetical function g. In the preceding chapter we found expressions of this form for $\zeta(s)\phi_s(n)/n^s$ and $\zeta(2s)\psi_s(n)/\zeta(s)n^s$.

Theorem 6.10. If, for each $\varepsilon > 0$, $g(r)/r^{\varepsilon}$ is bounded, then

$$\sum_{n\leq x} f(n) = Bx + O(1) ,$$

where

$$B = \sum_{r=1}^{\infty} \frac{g(r)\phi(r)}{r^{s+1}} .$$

Proof. If $\varepsilon > 0$ is small enough then $s - \varepsilon > 1$ and

$$\frac{|g(r)|}{r^s} = \frac{|g(r)|}{r^{\varepsilon}} \frac{1}{r^{s-\varepsilon}} \leq K \frac{1}{r^{s-\varepsilon}}$$

for some real number K. Hence the series for $f(n)$ converges absolutely for $s > 1$. Since $\phi(n) \leq n$ for all n, this shows also that the series for B converges absolutely for $s > 1$.

$$\sum_{n\leq x} f(n) = \sum_{n\leq x} \sum_{\substack{r=1 \\ (r,n)=1}}^{\infty} \frac{g(r)}{r^s}$$

$$= \sum_{r=1}^{\infty} \frac{g(r)}{r^s} \left(\sum_{\substack{n\leq x \\ (r,n)=1}} 1 \right),$$

and the rearrangement of the series if valid by absolute convergence. The sum in parentheses is $\phi(x,r)$: hence

$$\sum_{n\leq x} f(n) = \sum_{r=1}^{\infty} \frac{g(r)}{r^s} \left(\frac{\phi(r)x}{r} + h(r)\right)$$

$$= Bx + \sum_{r=1}^{\infty} \frac{g(r)h(r)}{r^s} ,$$

where h is a function such that $h(r)/\theta(r)$ is bounded. Let $\varepsilon > 0$ be so small that $s - 2\varepsilon > 1$. Since $g(r)/r^{\varepsilon}$ and $h(r)/\theta(r)$ are bounded, and by Exercise 5.79, $\tau(r)/r^{\varepsilon}$ is bounded, and since $\theta(n) \leq \tau(r)$ for all r, there is a real number K such that

$$\frac{|g(r)h(r)|}{r^{2\varepsilon}} \leq K \quad \text{for all } r .$$

Thus,

$$\frac{|g(r)h(r)|}{r^s} \leq K \frac{1}{r^{s-2\varepsilon}} ,$$

and consequently the series $\sum_{r=1}^{\infty} \frac{g(r)h(r)}{r^s}$ converges. □

It was shown in Chapter 5 that for all n and $s > 1$,

$$\frac{\phi_s(n)}{n^s} = \frac{1}{\zeta(s)} \sum_{\substack{r=1 \\ (r,n)=1}} \frac{1}{r^s} .$$

Therefore, for $s > 1$,

$$\sum_{n \leq x} \frac{\phi_s(n)}{n^s} = \frac{B_1 x}{\zeta(s)} + O(1) ,$$

where

$$B_1 = \sum_{r=1}^{\infty} \frac{\phi(r)}{r^{s+1}} = \frac{\zeta(s)}{\zeta(s+1)} .$$

Note that the bound in the term $O(1)$ depends on s . In particular, if $k \geq 2$,

$$\sum_{n \leq x} \frac{J_k(n)}{n^k} = \frac{x}{\zeta(k+1)} + O(1) .$$

It was also shown in Chapter 5 that for all n and $s > 1$,

$$\frac{\psi_s(n)}{n^s} = \frac{\zeta(s)}{\zeta(2s)} \sum_{\substack{r=1 \\ (r,n)=1}}^{\infty} \frac{\lambda(r)}{r^s} .$$

Since $\lambda(r)/r^{\varepsilon}$ is bounded for $\varepsilon > 0$,

$$\sum_{n \leq x} \frac{\psi_s(n)}{n^s} = \frac{\zeta(s)}{\zeta(2s)} B_2 x + O(1) ,$$

where

$$B_2 = \sum_{r=1}^{\infty} \frac{\lambda(r)\phi(r)}{r^{s+1}} = \frac{\zeta(2s)\zeta(s+1)}{\zeta(s)\zeta(2s+2)} \quad \text{for} \quad s > 1$$

by Exercise 5.22. Thus, for $s > 1$,

$$\sum_{n \leq x} \frac{\psi_s(n)}{n^s} = \frac{\zeta(s+1)}{\zeta(2s+2)} x + O(1) .$$

Exercises for Chapter 6

6.1. Let x and y be real numbers and m and n integers, $n > 0$.

(a) $[x + m] = [x] + m$

(b) $[x] + [-x] = 0$ if x is an integer, and $= -1$ otherwise.

(c) $[x + y] \leq [x] + [y]$.

(d) $\left[\frac{[x]}{n}\right] = \left[\frac{x}{n}\right]$.

(e) $[x] - 2[x/2] = 0$ or 1 .

(f) If $x \leq y$ there are $[y] - [x]$ integers k with $x < k \leq y$.

(g) If $x > 0$ there are $[x/n]$ positive multiples of n not exceeding x .

6.2. Suppose that $g(x)$ and $h(x)$ are positive for all large x .

(a) If $f_1(x) = O(g(x))$ and $f_2(x) = O(g(x))$ then $f_1(x) \pm f_2(x) = O(g(x))$.

(b) If $f(x) = O(g(x))$ then $f(x)^2 = O(g(x)^2)$.

(c) If $f(x) = O(g(x))$ and $g(x) = O(h(x))$, then $f(x) = O(h(x))$.

These properties of the big oh relation are used many times in the text without mention being made of that fact.

6.3. Suppose $g(x)$ and $h(g(x))$ are positive for all large x. In general it is not true that if $f(x) = O(g(x))$ then $h(f(x)) = O(h(g(x)))$. Give an example in which $f(x) = O(g(x))$, but $\log f(x) \neq O(\log g(x))$. Find some condition on $f(x)$ under which it is true that $f(x) = O(g(x))$ implies that $\log f(x) = O(\log g(x))$.

6.4. For $j = 1, 2, \ldots$ let $g_j(x)$ be positive for all large x, and suppose $f_j(x) = O(g_j(x))$. Let $h(x)$ be an increasing function of x. In general it is not true that

$$\sum_{j \le h(x)} f_j(x) = O\left(\sum_{j \le h(x)} g_j(x)\right)$$

(give an example). The relation is true, however, if there is a constant K that is independent of j such that $|f_j(x)| \le Kg_j(x)$ for all large x and all j.

6.5. Suppose that $f(x)$ is piecewise continuous and bounded for large x. If $s > 0$ then

$$\int_x^\infty \frac{f(t)}{t^{s+1}} \, dt = O\left(\frac{1}{x^s}\right) .$$

6.6. If C is Euler's constant then

$$C = 1 - \int_1^\infty \frac{t - [t]}{t^2} \, dt .$$

6.7. If $0 < s < 1$ then the improper integral

$$\int_1^\infty \frac{t - [t]}{t^{s+1}}\, dt$$

converges. Therefore, if $0 < s < 1$ then

$$\lim_{x\to\infty} \left(\sum_{n \leq x} \frac{1}{n^s} - \frac{x^{1-s}}{1-s} \right)$$

exists, and if we denote the limit by $\zeta(s)$, Proposition 6.3 holds when $0 < s < 1$.

6.8. If $s > 1$ then $\sum_{n>x} \frac{1}{n^s} = O(x^{1-s})$.

6.9. If $s \geq 0$ then

$$\sum_{n \leq x} n^s = \frac{x^{1+s}}{1+s} + O(x^s).$$

(Hint: let $f(x) = x^s$ in the Euler summation formula.)

6.10. If $s > 1$ then

$$\sum_{n>x} \frac{\mu(n)}{n^s} = O(x^{1-s}).$$

Therefore,

$$\sum_{n \leq x} \frac{\mu(n)}{n^s} = \frac{1}{\zeta(s)} + O(x^{1-s}).$$

6.11. Let $f(x)$ and $g(x)$ be functions of the real variable x for $x \geq 1$. Then

$$f(x) = \sum_{n \leq x} g(x/n) \quad \text{for all } x \geq 1$$

if and only if

$$g(x) = \sum_{n \leq x} \mu(n) f(x/n) \quad \text{for all } x \geq 1 .$$

6.12. Let $f(x)$ and $g(x)$ be functions of the real variable x for $x \geq 1$, and define $(f \circ h)(x)$ by

$$(f \circ g)(x) = \sum_{n \leq x} f(n) g(x/n) .$$

If $g(x) = 0$ when x is not an integer then $(f \circ g)(m) = (f * g)(m)$ for all integers m, and if $h(x)$ is any function of the real variable x then $(f \circ (g \circ h))(x) = ((f * g) \circ h)(x)$ for all $x \geq 1$.

6.13. (Generalization of Exercise 6.11.) Let $f(x)$ and $h(x)$ be functions of the real variable x for $x \geq 1$, and let g be an arithmetical function having an inverse. Then

$$f(x) = \sum_{n \leq x} g(n) h(x/n) \quad \text{for all } x \geq 1$$

if and only if

$$h(x) = \sum_{n \le x} g^{-1}(n) f(x/n) \quad \text{for all} \quad x \ge 1 .$$

Thus, if we set $g(x) = g^{-1}(x) = 0$ for x not an integer, then $f(x) = (g \circ h)(x)$ for all $x \ge 1$ if and only if $h(x) = (g^{-1} \circ f)(x)$ for all $x \ge 1$.

6.14. If f and g are arithmetical functions then for all $x \ge 1$,

$$\sum_{n \le x} (f * g)(n) = \sum_{d \le x} f(d) \sum_{n \le \frac{x}{d}} g(n) .$$

6.15. Let k be a positive integer and define the arithmetical function σ_{-k} by

$$\sigma_{-k}(n) = \sum_{d|n} \frac{1}{d^k} .$$

Then

$$\sum_{n \le x} \sigma_{-k}(n) = \zeta(k+1)\, x + \begin{cases} O(\log x) & \text{if } k = 1 \\ O(1) & \text{if } k \ge 2 . \end{cases}$$

6.16. $\displaystyle\sum_{n \le x} \frac{\tau(n)}{n} = \frac{1}{2} (\log x)^2 + 2C \log x + O(1) .$

6.17. $\displaystyle\sum_{n \le x} \frac{\phi(n)}{n} = \frac{6x}{\pi^2} + O(\log x).$

6.18. Let Φ_k be Klee's function, defined in Exercise 1.29. For $k \geq 2$,

$$\sum_{n \leq x} \Phi_k(n) = \frac{x^2}{2\zeta(2k)} + O(x) .$$

6.19. $\sum_{n \leq x} |\mu(n)| = \frac{6x}{\pi^2} + O(\sqrt{x})$ and $M(\mu^2) = \frac{6}{\pi^2}$.

6.20. If f is an arithmetical function such that $\sum_{n=1}^{\infty} f(n)/n$ converges, then $M(f) = 0$. (Hint: let $a_r = \sum_{n=r}^{\infty} f(n)/n$. The sequences $\{a_r\}$ and $\{\frac{1}{r}\sum_{j=1}^{r} a_j\}$ both converge to zero. Express $\sum_{n \leq x} f(n)$ in terms of the a_r's .)

6.21. Let g be an arithmetical function. If

$$\sum_{n=1}^{\infty} \frac{g(n)}{n}$$

converges absolutely, and if $f = g * \zeta$, then

$$M(f) = \sum_{n=1}^{\infty} \frac{g(n)}{n} .$$

6.22. Let k be a positive integer. If σ_{-k} is the function defined in Exercise 6.15 then $M(\sigma_{-k}) = \zeta(k + 1)$.

6.23. The series

$$\sum_{n=1}^{\infty} \frac{\mu(n)\phi(n)}{nJ_2(n)}$$

converges absolutely. (Hint: use Lemma 5.12.) In fact, it converges to A , the constant in Corollary 6.8. Furthermore,

$$6/\pi^2 < A < 1 .$$

6.24. If g is an arithmetical function then for $x \geq 1$,

$$\sum_{n\leq x} n\, g(n) = - \sum_{n\leq x} \sum_{k\leq n} g(k) + ([x] + 1)\sum_{n\leq x} g(n) .$$

6.25. For all n ,

$$\sum_{\substack{r\leq x \\ (r,n)=1}} |\mu(r)| = A_n x + O(\sqrt{x}\theta(n)) ,$$

where

$$A_n = \frac{6n\phi(n)}{\pi^2 J_2(n)} .$$

6.26. For all n ,

$$\sum_{\substack{r\leq x \\ (r,n)=1}} |\mu(r)|r = \frac{1}{2} A_n x^2 + O(x^{3/2}\theta(n)) .$$

6.27. Let g be a bounded arithmetical function and let the arithmetical function f be defined by

$$f(n) = \sum_{\substack{d|n \\ (d,n/d)=1}} d|\mu(d)|g(n/d) .$$

Then

$$\sum_{n\leq x} f(n) = \frac{3x^2}{\pi^2} \sum_{n=1}^{\infty} \frac{g(n)\phi(n)}{nJ_2(n)} + O(x^{3/2}) .$$

6.28. $\displaystyle\sum_{n\leq x} |\mu(n)|\phi(n) = \frac{3Ax^2}{\pi^2} + O(x^{3/2})$ (See Exercise 1.12 .)

6.29. $\displaystyle\sum_{n\leq x} \gamma(n) = \frac{1}{2} Ax^2 + O(x^{3/2})$.

6.30. For $s > 1$,

$$\sum_{n\leq x} \frac{n^s}{\phi_s(n)} = \zeta(s)D_1x + O(1) ,$$

where

$$D_1 = \prod_p \left(1 - \frac{(p-1)(p^{s-1}-1)}{p^{2s-1}}\right) .$$

6.31. For $s > 1$,

$$\sum_{n\leq x} \frac{n^s}{\psi_s(n)} = \frac{\zeta(2s)}{\zeta(s)} D_2 x + O(1) ,$$

where

$$D_2 = \prod_p \left(1 + \frac{(p-1)(p^{s-1}+1)}{p^{2s-1}}\right)$$

6.32. For $s > 1$,

$$\sum_{n\leq x} \frac{\phi_s(n)}{\psi_s(n)} = \frac{\zeta(2s)}{\zeta(s)^2} D_3 x + O(1) ,$$

where

$$D_3 = \prod_p \left(2\,\frac{1 - p^{-s}}{1 - p^{1-s}} - 1\right) .$$

6.33. Let S be a set of ordered pairs $\langle x,y \rangle$ of positive integers with $x \leq y$. For a positive integer n let

S_n = the number of $\langle x,y \rangle \in S$ such that $x \leq y \leq n$.

Since there are $n(n + 1)/2$ ordered pairs $\langle x,y \rangle$ of positive integers with $x \leq y \leq n$,

$$\lim_{n\to\infty} \frac{S_n}{\frac{1}{2}\,n(n+1)}$$

can be interpreted as the probability that if positive integers x and y are chosen at random, then $\langle x,y \rangle \in S$. If S is determined by some property P of pairs of integers, this is the probability that a pair of integers has property P. The probability that for a pair of integers x and y, $(x,y)_k = 1$, is $\frac{1}{\zeta(2k)}$. In particular, the probability that two positive integers chosen at random are relatively prime is $6/\pi^2$.

6.34. The probability that if two positive integers x and y are chosen at random it is true that $(x,y)_U = 1$, where x is the smaller of the two (i.e., $x \leq y$), is A, the constant of Corollary 6.8. Recall that by Exercise 6.23, $A > 6/\pi^2$.

Notes on Chapter 6

Many of the results in this chapter are classical. Some are more recent. Theorem 6.7 and the asymptotic formulas for the summatory functions of ϕ_U and σ_U in Corollaries 6.8 and 6.9 are due to E. Cohen [60c]. The formulas and other statements in Exercises 6.23, 6.25-6.29 and 6.34 are from the same paper. These results have been generalized by D. Suryanarayana [71c].

Theorem 6.10 was proved by E. Cohen [61d]. He gave many applications of the theorem, including those in the text and in Exercises 6.30-6.32.

The result in Exercise 6.20 is an old one. The proof suggested there and the result in Exercise 6.21 are from a paper by

J. G. van der Corput [39]. There has been much work done on the problem of the existence of mean values of arithmetical functions. A survey of the work done on this and related topics, with many references, was published by W. Schwarz [76].

The asymptotic formula for the summatory function of Klee's function in Exercise 6.18 was obtained by P. J. McCarthy [58]. His paper also contains the result in Exercise 6.33: the case $k = 1$ is classical, and the case $k = 2$ had been treated earlier by E. K. Haviland [44] and J. Christopher [56].

Chapter 7
Generalized Arithmetical Functions

Many of the properties of arithmetical functions, especially inversion properties and arithmetical identities, hold in a much more general setting than we have used in this book. In this chapter we shall introduce the reader to this general setting, look at various examples, and obtain some general results that can be applied in the special situations contained in the examples.

Let P be a partially ordered set, abbreviated poset. That is, P consists of a nonempty set, also denoted by P, and a relation $\leq$ on P such that for all $a, b, c \in P$,

(i) $a \leq a$,

(ii) if $a \leq b$ and $b \leq a$ then $a = b$ and

(iii) if $a \leq b$ and $b \leq c$ then $a \leq c$.

A relation on the set P having properties (i)-(iii) is called a partial ordering on P.

If $x, y \in P$ then the set

$$[x,y] = \{z : z \in P \text{ and } x \leq z \leq y\}$$

is called the interval determined by x and y. The poset P is said to be locally finite if $[x,y]$ is finite for all $x, y \in P$. The only posets of any interest to us are the locally finite ones, so we shall assume once and for all that P is locally finite. We must be careful to note in each example that the poset in that example is locally finite.

A complex-valued function f on $P \times P$ such that $f(x,y) = 0$ whenever $x \not\leq y$ is called an <u>incidence function</u> of P. For reasons that will be evident later, these might also be called <u>generalized arithmetical functions</u>, and this being a book on number theory, we think of them that way even though we call them incidence functions. Thus, the title of this chapter.

Let $F(P)$ denote the set of incidence functions of P. This set is certainly not empty: it contains δ and ζ where

$$\delta(x,y) = \begin{cases} 1 & \text{if } x = y \\ 0 & \text{if } x \neq y \end{cases}$$

and

$$\zeta(x,y) = \begin{cases} 1 & \text{if } x \leq y \\ 0 & \text{if } x \not\leq y \end{cases}$$

If f and g are in $F(P)$ their <u>sum</u> $f + g$ is defined by

$$(f + g)(x,y) = f(x,y) + g(x,y) \quad \text{for all } x, y \in P ,$$

their <u>product</u> fg by

$$(fg)(x,y) = f(x,y)g(x,y) \quad \text{for all } x, y \in P$$

and their <u>convolution</u> $f * g$ by

$$(f * g)(x,y) = \sum_{x \leq z \leq y} f(x,z)g(z,y) \quad \text{for all } x, y \in P ,$$

where a sum over an empty set is taken to be zero.

The binary operation of multiplication is commutative and associative, and distributes through addition. With respect to addition and convolution, $F(P)$ is a ring with unity element δ : it is not necessarily a commutative ring. (See Exercise 7.1.) The ring $F(P)$ is called the incidence ring of P, or the ring of generalized arithmetical functions defined on P.

An incidence function f of P is said to have an inverse if there is an incidence function g of P such that $f * g = g * f = \delta$. If there are two such functions, g and g', then

$$g = g * \delta = g * (f * g') = (g * f) * g' = \delta * g' = g' .$$

Thus, if a function f has an inverse, it is unique: it will be denoted by f^{-1}.

Proposition 7.1. An incidence function f of P has an inverse if and only if $f(x,x) \neq 0$ for all $x \in P$.

Proof. If f has an inverse then for all $x \in P$, $1 = \delta(x,x) = (f * f^{-1})(x,x) = f(x,x)f^{-1}(x,x)$: hence $f(x,x) \neq 0$.

Conversely, suppose $f(x,x) \neq 0$ for all $x \in P$. We shall define a function $g \in F(P)$ such that $g * f = \delta$. If $x \not\leq y$ let $g(x,y) = 0$. If $x \leq y$ the value of $g(x,y)$ will be defined by induction on $\#[x,y]$, the number of elements of $[x,y]$. If $x = y$ let

$$g(x,x) = \frac{1}{f(x,x)} .$$

Suppose that $x < y$ and that $g(u,v)$ has been defined for all $u, v \in P$

such that $u \leq v$ and $\#[u,v] < \#[x,y]$. If $x \leq z < y$ then $\#[x,z] < \#[x,y]$: hence $g(x,z)$ has been defined. Let

$$g(x,y) = -\frac{1}{f(y,y)} \sum_{x \leq z < y} g(x,z)f(z,y) .$$

For all $x \in P$, $g(x,x)f(x,x) = 1$, and if $x < y$,

$$\sum_{x \leq z \leq y} g(x,z)f(z,y) = 1 .$$

Thus, $g * f = \delta$.

Since $g(x,x) \neq 0$ for all $x \in G$, there is a function $h \in F(P)$ such that $h * g = \delta$. Then

$$f * g = \delta * (f * g) = (h * g) * (f * g) .$$

$$= (h * (g * f)) * g = (h * \delta) * g = h * g = \delta .$$

Therefore, g is an inverse of f . □

The function ζ has an inverse. It is called the <u>Möbius function</u> of P , and will be denoted by μ . If $f \in F(P)$ then

$$g = f * \zeta \quad \text{if and only if} \quad f = g * \mu ,$$

and

$$h = \zeta * f \quad \text{if and only if} \quad f = \mu * h .$$

If $x, y \in P$ then we say that y <u>covers</u> x if $x < y$ and if there are no elements z of P such that $x < z < y$. Suppose that y covers x . Then

$$0 = \delta(x,y) = (\mu * \zeta)(x,y) = \mu(x,x) + \mu(x,y) .$$

hence

$$\mu(x,y) = - \mu(x,x) = -1 .$$

If $x < y$ but y does not cover x , there is nothing we can say in general about the value of $\mu(x,y)$. However, there is one case in which we can determine $\mu(x,y)$ with little effort. Suppose that P is a <u>totally ordered set</u>, or <u>chain</u>, i.e., suppose that for all $x, y \in P$, either $x \leq y$ or $y \leq x$. If $x < y$ then the elements of $[x,y]$ can be indexed $z_0, z_1, \ldots, z_n$ so that

$$x = z_0 < z_1 < \ldots < z_n = y .$$

For $i = 0, 1, \ldots, n - 1$, z_{i+1} covers z_i : hence $\mu(z_i, z_{i+1}) = -1$. Assuming that y does not cover x , we have $n \geq 2$. Then

$$\mu(x,x) = 1 ,$$

$$\mu(x,z_1) = -1 ,$$

$$\mu(x, z_2) = - \mu(x,x) - \mu(x,z_1) = 0 ,$$

and by induction it follows that

$$\mu(x,z_i) = 0 \quad \text{for all } i \geq 2 .$$

In particular, $\mu(x,y) = 0$. Likewise, $\mu(z_{n-1},y) = -1$ and $\mu(z_i,y) = 0$ for $i \leq n - 2$.

For example, let N be the set of positive integers with the usual ordering. This poset is a locally finite chain and for all $n \in N$,

$$\mu(n,n) = 1 ,$$

$$\mu(n,n + 1) = -1 ,$$

$$\mu(n,m) = 0 \quad \text{for all} \quad m \geq n + 2 .$$

Elements x and y of P are said to have a <u>least upper bound</u> if there is an element u of P such that $x \leq u$ and $y \leq u$, and if $x \leq z$ and $y \leq z$ imply that $u \leq z$. If a least upper bound u exists it is unique, and will be denoted by $x \vee y$.

Elements x and y of P are said to have a <u>greatest lower bound</u> if there is an element v of P such that $v \leq x$ and $v \leq y$, and $z \leq x$ and $z \leq y$ imply that $z \leq v$. If a greatest lower bound v exists it is unique and will be denoted by $x \wedge y$.

The poset P is called a <u>lattice</u> if for all $x, y \in P$, x and y have a least upper bound and a greatest lower bound in P . This is a good place for a reminder that we consider only locally finite posets, and consequently only locally finite lattices. The next result, giving the fundamental properties of the binary operations $\langle x,y \rangle \mapsto x \vee y$ and $\langle x,y \rangle \mapsto x \wedge y$ on a lattice, is proved easily.

<u>Proposition 7.2</u>. Let L be a lattice. Then:

(1) $x \vee x = x \wedge x = x$ for all $x \in L$.

(2) $x \vee y = y \vee x$ and $x \wedge y = y \wedge x$ for all $x, y \in L$.

(3) $(x \vee y) \vee z = x \vee (y \vee z)$ and $(x \wedge y) \wedge z = x \wedge (y \wedge z)$ for all $x, y, z \in L$.

(4) $x \vee (x \wedge y) = x \wedge (x \vee y) = x$ for all $x, y \in L$.

(5) For $x, y \in L$ the statements $x \wedge y = x$, $x \vee y = y$ and $x \leq y$ are equivalent.

Properties (1)-(4) are characteristic of lattices in the sense described in Exercise 7.14.

A lattice L is called <u>distributive</u> if

$$x \wedge (y \vee z) = (x \wedge y) \vee (x \wedge z) \quad \text{for all} \quad x, y, z \in L$$

(see Exercises 7.15 and 7.16). If P is a chain then P is obviously a lattice, and it is equally obvious that P is distributive. On the other hand, the lattice L , given as in Exercise 7.4 by

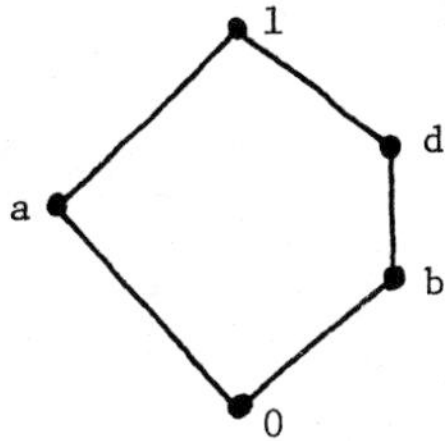

is not distributive, for

$$d \wedge (a \vee b) = d \wedge 1 = d, \text{ but } (d \wedge a) \vee (d \wedge b) = 0 \vee b = b .$$

On the set of positive integers the relation $m|n$ is a partial ordering. The resulting poset will be denoted by N_D . If $m|n$ then

$$[m,n] = \{d : d|n \text{ and } m|d\} :$$

hence N_D is locally finite. The poset N_D is a lattice, in which the least upper bound of two integers m and n is their least common multiple, and the greatest lower bound of m and n is their greatest common divisor. This lattice is distributive.

If f is an arithmetical function then we can associate with f an incidence function $\bar{f}$ of N_D defined by

$$\bar{f}(m,n) = \begin{cases} f(n/m) & \text{if } m|n \\ 0 & \text{if } m\nmid n . \end{cases}$$

Since $1|n$ for all n, the mapping $f \mapsto \bar{f}$ if one-one. Furthermore, it preserves addition and convolution, i.e., if f and g are arithmetical functions then $\overline{f+g} = \bar{f} + \bar{g}$ and $\overline{f * g} = \bar{f} * \bar{g}$, where the $*$ on the left is Dirichlet convolution and the $*$ on the right is the convolution in $F(N_D)$.[1] The former equality is clear, and the latter we shall verify. If $m|n$ then

$$(\bar{f} * \bar{g})(m,n) = \sum_{\substack{d|n \\ m|d}} \bar{f}(m,d)\bar{g}(d,n) = \sum_{\substack{d|n \\ m|d}} f(d/m)g(n/d)$$

$$= \sum_{e|\frac{n}{m}} f(e)g\left(\frac{n/m}{e}\right) = (f * g)(n/m) .$$

To generalize the preceding example, let A be a regular arithmetical

[1] In the language of abstract algebra, there is an injective homomorphism from the ring of arithmetical functions into the ring $F(N_D)$.

convolution. The relation $\leq$ on the set of positive integers defined by

$$m \leq n \qquad \text{if} \qquad m \in A(n)$$

is a partial ordering. The resulting locally finite poset will be denoted by N_A . If f is an incidence function of N_A then f is an incidence function of N_D . Furthermore, if $f, g \in F(N_A)$ then their sum in $F(N_A)$ is the same as their sum in $F(N_D)$, and their convolution in $F(N_A)$ is the same as their convolution in $F(N_D)$. This is because the extra terms that appear to occur in

$$\sum_{\substack{d|n \\ m|d}} f(m,d)g(d,n)$$

are actually equal to zero. Thus, $F(N_A)$ can be considered to be a subring of $F(N_D)$. One must be careful, however. The element ζ in $F(N_A)$ is not the same as the element ζ in $F(N_D)$, unless $A = D$.

With each arithmetical function f we can associate an incidence function $\tilde{f}$ of N_A defined by

$$\tilde{f}(m,n) = \begin{cases} f(n/m) & \text{if } m \in A(n) \\ 0 & \text{if } m \notin A(n) . \end{cases}$$

The mapping $f \mapsto \tilde{f}$ is one-one and if f and g are arithmetical functions and $m \in A(n)$ then

$$(\tilde{f} + \tilde{g})(m,n) = \tilde{f}(m,n) + \tilde{g}(m,n) = f(n/m) + g(n/m)$$

$$= (f + g)(n/m) = \widetilde{(f + g)}(m,n) .$$

and

$$(\tilde{f} * \tilde{g})(m,n) = (f *_A g)(n/m) = \widetilde{(f *_A g)}(m,n) .$$

Note that unless $A = D$, if f is an arithmetical function, then $\tilde{f} \in F(N_A)$ and $\bar{f} \in F(N_D)$ are not necessarily the same.

The poset N_A is not always a lattice. Since N_A is locally finite and since $1 \leq m$ and $1 \leq n$ for all m and n , every pair of elements of N_A has a greatest lower bound in N_A . However, if p is a prime and $\alpha \neq \beta$, then p^{α} and p^{β} have no upper bound at all in N_U . (In connection with this, see Exercise 7.19.) It is true, and this is obvious, that every nonempty interval of N_A is a lattice.

A poset P is called a <u>local lattice</u> if every nonempty interval in P is a lattice with respect to the partial ordering that it inherits from P . Every lattice L is a local lattice and if x,y belong to some interval in L then $x \vee y$ is their least upper bound in the interval and $x \wedge y$ is their greatest lower bound in the interval. A local lattice P is said to be <u>locally distributive</u>, or <u>locally modular</u>, if every nonempty interval in P is a distributive lattice, or a modular lattice (see Exercise 7.17). Note that if a poset P has an element a such that $a \leq x$ for all $x \in P$ then, because intervals are finite, every pair of elements x and y of P have a greatest lower bound $x \wedge y$ in P . A poset having this property is called a <u>lower semi-lattice</u>.

<u>Proposition 7.3</u>. If A is a regular arithmetical convolution then N_A is a lower semi-lattice and a locally distributive local lattice.

Proof. We refer to the properties (1')-(4') which define regular arithmetical convolutions, and which are stated immediately preceding Theorem 4.1. Because of (3') all that we need to show is that for every n, the interval $[1,n]$, i.e., $A(n)$, is a distributive lattice. In fact, if x and y are in an interval $[m,n]$, their least upper bound and greatest lower bound in that interval are exactly the same integers as their least upper bound and greatest lower bound in $[1,n]$.

Let $n = p_1^{\alpha_1} \cdots p_s^{\alpha_s}$ and $t_i = t_A(p_i^{\alpha_i})$ for $i = 1, \ldots, s$. If $x, y \in A(n)$ then by Theorem 4.1,

$$x = \prod_{i=1}^{s} p_i^{h_i t_i}, \quad 0 \le h_i \le \alpha_i/t_i ,$$

$$y = \prod_{i=1}^{s} p_i^{j_i t_i}, \quad 0 \le j_i \le \alpha_i/t_i ,$$

and by Corollary 4.2,

$$x \wedge y = \prod_{i=1}^{s} p_i^{t_i \min(h_i, j_i)}, \quad x \vee y = \prod_{i=1}^{s} p_i^{t_i \max(h_i, j_i)} .$$

Thus, $x \wedge y$ and $x \vee y$ are, respectively, nothing more or less than the greatest common divisor and least common multiple of the integers x and y. Therefore, for all $x, y, z \in A(n)$, $x \wedge (y \vee z) = (x \wedge y) \vee (x \wedge z)$. □

In the theory of arithmetical functions the notion of a multiplicative function plays a central role. It is natural to ask if a similar notion can be introduced into the study of incidence functions. We shall discuss one way of doing this.

Let L be a local lattice. An incidence function f of L is called <u>factorable</u> if f has an inverse and if

$$f(a \vee b,\ c \vee d) = f(a,c)f(b,d)$$

for all $a, b, c, d \in L$ such that a, b, c, d belong to some interval in L, and

$$c \wedge d \leq a \leq c,\ c \wedge d \leq b \leq d$$

(which is the same as saying that $a \leq c$, $b \leq d$ and $a \wedge b = c \wedge d$). If f is factorable then $f(a,a) = 1$ for all $a \in L$. The function ζ in $F(L)$ is factorable, but the function δ is not always factorable. For example, if L is the second lattice in Exercise 7.4, then $\delta(a \vee b,\ a \vee d) = 1$ but $\delta(a,a)\delta(b,d) = 0$.

The notion of factorable function does have one property that is certainly desirable. If f is an arithmetical function then by Exercise 1.10, f is a multiplicative function if and only if the associated function $\bar{f} \in F(N_D)$ is factorable.

It is not always true that inverses and convolutions of factorable functions are factorable. For example, consider once more the second lattice in Exercise 7.4. Since $\mu(a \vee b,\ a \vee d) = 1$ and $\mu(a,a)\mu(b,d) = -1$, the function μ is not factorable. Also, since $(\zeta * \zeta)(a \vee b,\ a \vee d) = 1$ and $(\zeta * \zeta)(a,a)(\zeta * \zeta)(b,d) = 2$, the function $\zeta * \zeta$ is not factorable.

The lattice in this example is not distributive and, as we shall see, distributivity is the key. (See also Exercise 7.22.)

Proposition 7.4. If L is a locally distributive local lattice then the inverse of a factorable function f in $F(L)$ is factorable.

Proof. Let a, b, c, d be elements of L belonging to some interval in L and such that $c \wedge d \leq a \leq c$ and $c \wedge d \leq b \leq d$. By Exercise 7.21 the mapping $\eta : [a,c] \times [b,d] \to [a \vee b, c \vee d]$ is one-one and onto. Thus, if $a \vee b = c \vee d$ then $a = c$ and $b = d$, and the required equality involving f^{-1} holds.

We proceed by induction. The induction hypothesis is this: $a \vee b \neq c \vee d$ and for all $a', b', c', d' \in L$ with $c' \wedge d' \leq a' \leq c'$ and $c' \wedge d' \leq b' \leq d'$, and for which $\#[a' \vee b', c' \vee d'] < \#[a \vee b, c \vee d]$,

$$f^{-1}(a' \vee b', c' \vee d') = f^{-1}(a', c')f^{-1}(b', d') .$$

Then,

$$f^{-1}(a \vee b, c \vee d) = - \sum_{a \vee b \leq z < c \vee d} f^{-1}(a \vee b, z)f(z, c \vee d)$$

$$= - \sum_{\substack{\langle x,y \rangle \in [a,c]\times[b,d] \\ \langle x,y \rangle \neq \langle c,d \rangle}} f^{-1}(a \vee b, x \vee y)f(x \vee y, c \vee d)$$

$$= - \sum_{\substack{a \leq x \leq c \\ b \leq y \leq d \\ \langle x,y \rangle \neq \langle c,d \rangle}} f^{-1}(a,x)f^{-1}(b,y)f(x,c)f(y,d) ,$$

by the induction hypothesis, and because f is factorable. Note that for every x and y, $x \vee y \neq c \vee d$ since η is one-one: hence $\#[a \vee b, x \vee y] < \#[a \vee b, c \vee d]$. Also note that $x \wedge y \leq a \leq x$ and $x \wedge y \leq b \leq y$, and $c \wedge d \leq x \leq c$ and $c \wedge d \leq y \leq d$.

To continue, the sum above is equal to

$$- f^{-1}(a,c) \sum_{b \leq y < d} f^{-1}(b,y)f(y,d) - f^{-1}(b,d) \sum_{a \leq x < c} f^{-1}(a,x)f(x,c)$$

$$- \left(- \sum_{a \leq x < c} f^{-1}(a,x)f(x,c) \right) \left(- \sum_{b \leq y < d} f^{-1}(b,y)f(y,d) \right)$$

$$= f^{-1}(a,c)f^{-1}(b,d) + f^{-1}(a,c)f^{-1}(b,d) - f^{-1}(a,c)f^{-1}(b,d)$$

$$= f^{-1}(a,c)f^{-1}(b,d) .$$

Therefore, f^{-1} is factorable. □

It is not surprising that this proof is essentially the same as the proof of Proposition 1.5.

Proposition 7.5. If L is a locally distributive local lattice then the convolution of two factorable functions f and g in $F(L)$ is factorable.

Proof. Let a, b, c, d be as in the proof of Proposition 7.4. Then

$$(f * g)(a,c)(f * g)(b,d)$$

$$= \left(\sum_{a \leq x \leq c} f(a,x)g(x,c) \right) \left(\sum_{b \leq y \leq d} f(b,y)g(y,d) \right)$$

$$= \sum_{\langle x,y \rangle \in [a,c]\times[b,d]} f(a,x)f(b,y)g(x,c)g(y,d)$$

$$= \sum_{\langle x,y \rangle \in [a,c]\times[b,d]} f(a \vee b,\ x \vee y)g(x \vee y,\ c \vee d)$$

$$= \sum_{a\ b \le z \le c\ d} f(a \vee b,\ z)g(z,\ c \vee d) = (f * g)(a \vee b,\ c \vee d)$$

Therefore, $f * g$ is factorable. □

The converses of Propositions 7.4 and 7.5 are true. In fact, a much weaker statement is true.

<u>Proposition 7.6</u>. Let L be a local lattice.

(a) If μ is factorable then L is locally distributive.

(b) If $\zeta * \zeta$ is factorable then L is locally distributive.

Outline of the proof. We shall give only an outline because the proof depends on a characterization of distributive lattices that is not properly a part of the presentation we are making. Suppose that L is not locally distributive, i.e., there are elements $x, y \in L$ with $x \le y$ such that the interval $[x,y]$, which is a lattice, is not distributive. Then one of the following is true:

(1) $[x,y]$ contains (distinct) elements $a, b, d, 0, 1$ such that $0 < a < 1$, $0 < b < d < 1$, $a \wedge b = a \wedge d = 0$ and $a \vee b = a \vee d = 1$.

(2) $[x,y]$ contains (distinct) elements $a, b, c, 0, 1$ such that

no two of a, b, c are comparable, $a \vee b = b \vee c = a \vee c = 1$ and $a \wedge b = b \wedge c = a \wedge c = 0$.[1)]

Suppose that (1) holds. If there is an element e in the interval such that $b < e < d$, then $a, b, e, 0, 1$ have the same properties as $a, b, d, 0, 1$. Thus we can assume that d covers b . Then $\mu(a \vee b, a \vee d) = 1$ and $\mu(a,a)\mu(b,d) = -1$: hence μ is not factorable. Also, $(\zeta * \zeta)(a \vee b, a \vee d) = 1$ and $(\zeta * \zeta)(a,a)(\zeta * \zeta)(b,d) = 2$: hence $\zeta * \zeta$ is not factorable.

The case in which (2) holds is left as Exercise 7.26. □

The next theorem is an immediate consequence of Propositions 7.4, 7.5 and 7.6.

Theorem 7.7. Let L be a local lattice. Inverses and convolutions of factorable functions in $F(L)$ are factorable [2)] if and only if L is locally distributive.

Let A be a regular arithmetical convolution. If f is a multiplicative function and $\tilde{f}$ is the associated incidence function in $F(N_A)$, then by what was shown in the proof of Proposition 7.3, $\tilde{f}$ is factorable. If f^{-1} is the inverse of f with respect to the convolution A then $\tilde{f} * \widetilde{f^{-1}} = \widetilde{f *_A f^{-1}} = \delta$: thus $\widetilde{f^{-1}} = \tilde{f}^{-1}$, which is factorable by Theorem 7.7. It follows that f^{-1} is multiplicative. A more general result is in Exercise 4.4.

---

1) In the language of lattice theory, if a lattice is not distributive then it has a sublattice isomorphic to one of the lattices in Exercise 7.4. See [C], p. 66 and p. 69.

2) I.e., the factorable functions form a subgroup of the group of units of the ring $F(L)$.

Let L be a local lattice. Among the incidence functions of L are ones which are analogues of familiar arithmetical functions. Let H be a function from L into the set of positive integers. Assume that if x and y belong to an interval in L then $H(x \wedge y)$ is the greatest common divisor of $H(x)$ and $H(y)$, and $H(x \vee y)$ is then least common multiple. Let k be a nonegative integer and define the incidence function $\nu_{k,H}$ of L by

$$\nu_{k,H}(x,y) = \begin{cases} \dfrac{H(y)^k}{H(x)^k} & \text{if } x \le y \\ 0 & \text{if } x \not\le y . \end{cases}$$

Since $\nu_{k,H}(x,x) = 1$ for all $x \in L$, $\nu_{k,H}$ has an inverse. For every choice of H,

$$\nu_{0,H} = \zeta \quad \text{and} \quad \nu_{0,H}^{-1} = \mu .$$

If $x \le z \le y$ then

$$\nu_{k,H}(x,z)\nu_{k,H}(z,y) = \frac{H(z)^k}{H(x)^k}\,\frac{H(y)^k}{H(z)^k} = \frac{H(y)^k}{H(x)^k} = \nu_{k,H}(x,y) .$$

Thus, $\nu_{k,H}$ is completely factorable, and consequently $\nu_{k,H}^{-1} = \mu\nu_{k,H}$ (see Exercise 7.13). Furthermore, if a and b belong to an interval in L then

$$\nu_{k,H}(a \wedge b,\ a) = \left(\frac{H(a)}{H(a\wedge b)}\right)^k = \left(\frac{H(a\vee b)}{H(b)}\right)^k = \nu_{k,H}(b,\ a \vee b) ,$$

i.e., (b) of Exercise 7.24 holds. Therefore, by Exercise 7.25, $\nu_{k,H}$ is factorable.

The analogues of the arithmetical functions σ_k and J_k are the incidence functions $\sigma_{k,H} = \nu_{k,H} * \zeta$ and $J_{k,H} = \nu_{k,H} * \mu$. If $L = N_D$ and if $H(n) = n$ for all n, then $\sigma_{k,H} = \bar{\sigma}_k$ and $J_{k,H} = \bar{J}_k$. By Theroem 7.7, $\sigma_{k,H}$ and $J_{k,H}$, and their inverses, are factorable if the local lattice L is locally distributive.

From the point of view of this book one of the interesting applications of incidence functions is to arithmetical identities. The rest of this chapter will be devoted to a discussion of some very general identities. Before stating the identities we shall do a little preparatory work.

Let L be a locally distributive local lattice in which every pair of elements has a greatest lower bound in L, and let f and g be factorable functions in $F(L)$. Let $t \in L$ and for every $x, y \in L$ with $x \leq y \wedge t$ set

$$h(x,y) = \sum_{\substack{x \leq z \leq y \\ z \wedge t = x}} f(x,z)g(z,y) .$$

Note that h is not necessarily an incidence function of L because $h(x,y)$ may not be defined for all $x, y \in L$. However, it is true that if a, b, c, d belong to some interval in L and if $a \leq c \wedge t$, $b \leq d \wedge t$ and

$$c \wedge d \leq a \leq c \quad , \quad c \wedge d \leq b \leq d ,$$

then

$$h(a \vee b,\ c \vee d) = h(a,c)h(b,d)\ .$$

The proof of this assertion is similar to that of Proposition 7.5, and it makes use of Exercise 7.21. We observe that

$$a \vee b \leq (c \wedge t) \vee (d \wedge t) = (c \vee d) \wedge t$$

since L is locally distributive. Then

$$h(a \vee b,\ c \vee d) = \sum_{\substack{a\vee b\leq z\leq c\vee d \\ z\wedge t=a\vee b}} f(a \vee b,\ z)g(z,\ c \vee d)$$

$$= \sum_{\substack{\langle x,y \rangle\in[a,c]\times[b,d] \\ (x\vee y)\wedge t=a\vee b}} f(a \vee b,\ x \vee y)g(x \vee y,\ c \vee d)\ .$$

If η is the mapping of Exercise 7.21, then since

$$\eta(\langle x \wedge t,\ y \wedge t \rangle) = (x \wedge t) \vee (y \wedge t)$$

$$= (x \vee y) \wedge t = a \vee b = \eta(\langle a,b \rangle)\ ,$$

and η is one-one (and, of course, $a \leq x \wedge t \leq c$ and $b \leq y \wedge t \leq d$), $x \wedge t = a$ and $y \wedge t = b$. Thus

$$h(a \vee b,\ c \vee d) = \sum_{\substack{a\leq x\leq c \\ b\leq y\leq d \\ x\wedge t=a \\ y\wedge t=b}} f(a,x)f(b,y)g(x,c)g(y,d)$$

$$= \left(\sum_{\substack{a \le x \le c \\ x \wedge t = a}} f(a,x)g(x,c) \right) \left(\sum_{\substack{b \le y \le d \\ y \wedge t = b}} g(b,y)g(y,d) \right)$$

$$= h(a,c)h(b,d) \ .$$

We assume for the rest of this chapter that in addition to the conditions already imposed on L, each of its intervals is a product of chains. See Exercises 7.11 and 7.21 for the product of two lattices: the extension to products of more than two lattices is immediate. Since L is locally finite, for each interval there is only a finite number of chains involved in the product and each of them is finite.

Let $f \in F(L)$. A function $h \in F(L)$ is called a _companion function_ for f if $h(x,y) = f(x,y) - 1$ whenever y covers x. By Exercise 7.31, if f is factorable then it has a factorable companion function.

The function f is said to _constant on chains_ if for all $x, y \in L$ such that $x \le y$ and $[x,y] = \{x_0, x_1, \ldots, x_n\}$ is a chain,

$$x = x_0 < x_1 < \ldots < x_n = y \ ,$$

we have $f(x,y) = f(x_i, x_j)$ for all i, j with $0 \le i < j \le n$.

If A is a regular arithmetical convolution then N_A satisfies all the conditions that have been imposed on L. If $m \in A(n)$ and if

$$n = \prod_{i=1}^{s} p_i^{\alpha_i} \quad , \qquad m = \prod_{i=1}^{s} p_i^{\beta_i} \ ,$$

then the interval $[m,n]$ is the product of the intervals $[p_i^{\alpha_i}, p_i^{\beta_i}]$ for $i = 1, \ldots, s$, and each of these is a chain. If $f(n) = \phi(n)/n$ for all n then the associated function $\overline{f} \in F(N_D)$ is constant on chains.

Theorem 7.8. Let f be a factorable function in $F(L)$, constant on chains, and let h be a factorable companion function for f. If $x, y, t \in L$ and $x \leq y \wedge t$ then

$$\sum_{\substack{x \leq z \leq y \\ z \wedge t = x}} f(x,z)\mu(z,y) = \mu(x,y)\mu(y \wedge t, y)h(y \wedge t, y) .$$

Proof. We can assume that $x \leq t \leq y$, for if $t' = y \wedge t$ and $x \leq z \leq y$, then $z \wedge t = z \wedge t'$. Thus, if $g(x,y)$ is the sum on the left, we want to show that

$$g(x,y) = \mu(x,y)\mu(t,y)h(t,y) .$$

We shall show that this holds for all x, y, t with $x \leq t \leq y$ if it holds whenever $[x,y]$ is a chain. Assuming for the moment that this is true, we show now that the equality does hold when $[x,y]$ is a chain. If $x = y$ both sides are equal to one. Suppose that $x < y$ and that $[x,y] = \{z_0, z_1, \ldots, z_n\}$, where $x = z_0 < z_1 < \ldots < z_n = y$. Suppose further that $t = z_j$.

If $j = 0$ then since $\mu(z_i, y) = 0$ for $i \leq n - 2$,

$$g(x,y) = f(x,z_{n-1})\mu(z_{n-1}, y) + f(x,y)\mu(y,y)$$

$$= - f(x,z_{n-1}) + f(x,y)$$

$$= \begin{cases} -1 + f(x,z_1) & \text{if } n = 1 \\ - f(x,z_1) + f(x,z_1) = 0 & \text{if } n \geq 2 \end{cases}$$

and

$$\mu(x,y)\mu(t,y)h(t,y)$$

$$= \mu(x,y)\mu(x,y)h(x,y) = \begin{cases} (-1)^2(f(x,z_1) - 1) & \text{if } n = 1 \\ 0 & \text{if } n \geq 2 . \end{cases}$$

If $j > 0$ then

$$g(x,y) = f(x,x)\mu(x,y) = \begin{cases} -1 & \text{if } n = 1 \\ 0 & \text{if } n \geq 2 \end{cases}$$

and

$$\mu(x,y)\mu(t,y)h(t,y) = \begin{cases} \mu(x,y)\mu(y,y)h(y,y) = -1 & \text{if } n = 1 \\ 0 & \text{if } n \geq 2 . \end{cases}$$

Now consider the case when $[x,y]$ is the product of two chains (the case when $[x,y]$ is the product of more than two chains can be treated in an exactly similar way). Let $[x,y] = L_1 \times L_2$, where

$$L_1 = \{w_0, w_1, \ldots, w_m\} , \; w_0 < w_1 < \ldots < w_m ,$$

$$L_2 = \{z_0, z_1, \ldots, z_n\} , \; z_0 < z_1 < \ldots < z_n .$$

If we identify each w_i with $\langle w_i, z_0\rangle$ and each z_i with $\langle w_0, z_i\rangle$

then L_1 and L_2 can be considered to be subsets of $[x,y]$ with the ordering inherited from $[x,y]$. Furthermore, $L_1 = [w_0, w_m]$ and $L_2 = [z_0, z_n]$. Since $w_0 \leq w_n \wedge t$, $z_0 \leq z_n \wedge t$ and

$$x = w_m \wedge z_n \leq w_0 \leq w_m , \; x = w_m \wedge z_n \leq z_0 \leq z_n ,$$

we have

$$\sum_{\substack{x \leq z \leq y \\ z \wedge t = x}} f(x,z)\mu(z,y)$$

$$= \left(\sum_{\substack{w_0 \leq w \leq w_m \\ w \wedge t = w_0}} f(x_0,w)\mu(w,w_m) \right) \left(\sum_{\substack{z_0 \leq z \leq z_n \\ z \wedge t = z_0}} f(z_0,z)\mu(z,z_n) \right) .$$

On the other hand, since h is factorable,

$$h(y \wedge t,y) = h((w_m \vee z_n) \wedge t \; , \; w_m \vee z_n)$$

$$= h((w_m \wedge t) \vee (z_n \wedge t), \; w_m \vee z_n)$$

$$= h(w_m \wedge t, \; w_m)h(z_n \wedge t, \; z_n) \; ,$$

and a similar equality holds for $\mu(x,y)$ and $\mu(y \wedge t,y)$. This completes the proof of the theorem. □

If we apply Theorem 7.8 to the function $\bar{f} \in F(N_D)$ associated with a multiplicative function f we obtain the following result.

Corollary 7.9. Let f be a multiplicative function such that $f(p^\alpha) = f(p)$ for all primes p and all $\alpha \geq 1$. Let h be a multiplicative function such that $h(p) = f(p) - 1$ for every prime p. Then for all n and r,

$$\sum_{\substack{d|r \\ (n,d)=1}} f(d)\mu(r/d) = \mu(r)\mu(m)h(m) ,$$

where $m = r/(n,r)$.

The only requirement that h must satisfy concerns $h(p)$ for every prime p : the values of $h(p^\alpha)$ for $\alpha > 1$ can be chosen arbitrarily. One choice for h is $f * \mu$, but h does not have to be this funciton.

For example, suppose that $f(r) = r/\phi(r)$ for all r . Then $f(p^\alpha) = p/(p-1)$ for all primes p and all $\alpha \geq 1$. If $h(r) = 1/\phi(r)$ for all r then for every prime p ,

$$h(p) = \frac{1}{p - 1} = f(p) - 1 .$$

Therefore, for all n and r ,

$$\sum_{\substack{d|r \\ (n,d)=1}} \frac{d}{\phi(d)} \mu(r/d) = \frac{\mu(r)\mu(m)}{\phi(m)} ,$$

which is the Brauer-Rademacher identity obtained as a special case of Theorem 2.5.

If θ is the function defined in Exercise 1.24 then $\theta(p^\alpha) = 2$ for all primes p and all $\alpha \geq 1$. Taking $h = \zeta$, we find that for all n and r ,

$$\sum_{\substack{d|r \\ (n,d)=1}} \theta(d)\mu(r/d) = \mu(r)\mu(m) .$$

Exercises for Chapter 7

7.1. Let P be a poset. With respect to addition and convolution, $F(P)$ is a ring, not necessarily commutative, and with unity element δ.

7.2. In the proof of Proposition 7.1 we could have defined g as follows:

$$g(x,x) = \frac{1}{f(x,x)} \quad \text{for all} \quad x \in P ,$$

and if $x < y$,

$$g(x,y) = -\frac{1}{f(x,x)} \sum_{x<z\leq y} f(x,z)g(z,y) .$$

7.3. Let P be the poset represented by the diagram on the right ($x < y$ if there is a polygonal path going up from x to y). There is an incidence function f of P such that $f * \zeta \neq \zeta * f$. Determine the Möbius function of P.

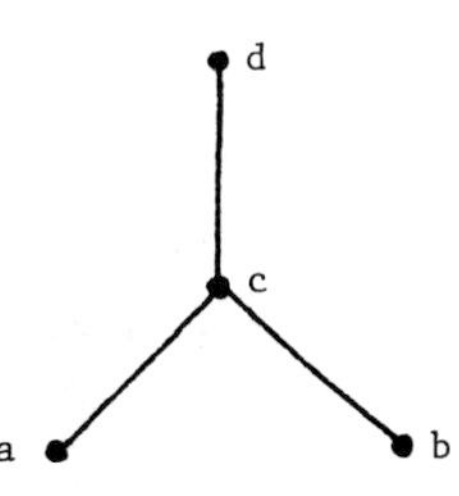

7.4. Determine the Möbius functions of the posets

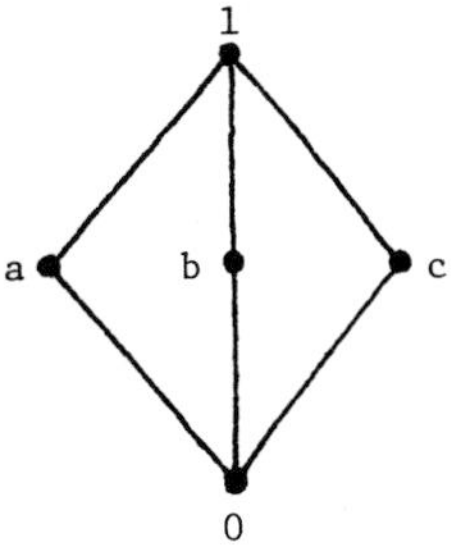

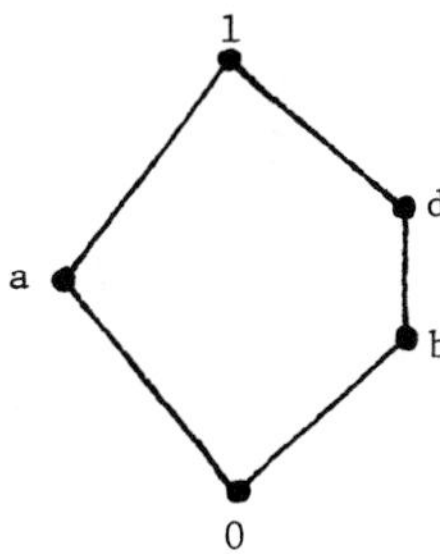

7.5. If μ is the Möbius function of a poset P and if $x, y \in P$ and $x < y$, then

$$\mu(x,y) = -\sum_{x \leq z < y} \mu(x,z) = -\sum_{x < z \leq y} \mu(z,y) .$$

7.6. Let P be a poset and let f be a complex-valued function on P such that for some $a \in P$, $f(x) = 0$ unless $x \geq a$. If

$$g(x) = \sum_{z \leq x} f(z) \quad \text{for all} \quad x \in P$$

then

$$f(x) = \sum_{z \leq x} g(z)\mu(z,x) \quad \text{for all} \quad x \in P ,$$

and conversely.

7.7. Let P be a poset and let f be a complex-valued function on P such that for some $b \in P$, $f(x) = 0$ unless $x \leq b$. If

$$g(x) = \sum_{x \leq z} f(z) \quad \text{for all} \quad x \in P$$

then

$$f(x) = \sum_{x \leq z} \mu(x,z)g(z) \quad \text{for all} \quad x \in P ,$$

and conversely.

7.8. Let P' be a sub-poset of a poset P, i.e., $P' \subseteq P$ and $\leq$ agrees with $\leq'$ on $P' \times P'$. If the interval $[x,y]$ in P is contained in P' then $\mu'(x,y) = \mu(x,y)$. If the interval $[x,y]$ in P is not contained in P', it can happen that $\mu'(x,y) \neq \mu(x,y)$.

7.9. Suppose that P and P' are isomorphic posets, i.e., there is a one-one mapping f from P onto P' such that if $x, y \in P$ then $x \leq y$ if and only if $f(x) \leq' f(y)$. Then $\mu'(f(x), f(y)) = \mu(x,y)$ for all $x, y \in P$.

7.10. Let S be a nonempty set and let P be the poset consisting of the finite subsets of S, ordered by set inclusion. If $X, Y \in P$ and $X \subseteq Y$, then

$$\mu(X,Y) = (-1)^{\#Y - \#X} .$$

(Prove this by using the definition of the Möbius function.)

7.11. Let P_1 and P_2 be posets. Let the set P be the Cartesian product of P_1 and P_2 and define $\leq$ on $P \times P$ by

$$\langle x_1, x_2\rangle \leq \langle y_1, y_2\rangle \text{ if } x_1 \leq y_1 \text{ and } x_2 \leq y_2 .$$

Then $\leq$ is a partial ordering on P. For all $\langle x_1, x_2\rangle, \langle y_1, y_2\rangle \in P$,

$$\mu(\langle x_1, x_2\rangle, \langle y_1, y_2\rangle) = \mu_1(x_1, y_1)\mu_2(x_2, y_2) .$$

7.12. The result of the preceding exercise can be extended readily to the product of several posets. Apply it to the product of several copies of the poset $\{0,1\}$, with $0 < 1$, to obtain an alternate proof of the assertion in Exercise 7.10.

7.13. Let P be a poset. An incidence function f of P is said to be completely factorable if $f(x,y) = f(x,z)f(z,y)$ for all $x, y, z \in P$ for which $x \leq z \leq y$. The function f is completely factorable if and only if $f(g * h) = fg * fh$ for all $g, h \in F(P)$. If f is completely factorable and has an inverse, then $f^{-1} = \mu f$.

7.14. Let L be a nonempty set with two binary operations $\langle x,y \rangle \mapsto x \vee y$ and $\langle x,y \rangle \mapsto x \wedge y$ that satisfy (1)-(4) of Proposition 7.2. There is a unique partial ordering $\leq$ on L such that L together with $\leq$ is a lattice, and such that for all $x, y \in L$, $x \vee y$ is the least upper bound and $x \wedge y$ the greatest lower bound of x and y in the lattice L.

7.15. A lattice L is distributive if and only if

$$x \vee (y \wedge z) = (x \vee y) \wedge (x \vee z) \quad \text{for all } x, y, z \in L .$$

7.16. A lattice L is distributive if and only if

$$(x \vee y) \wedge (x \vee z) \wedge (y \vee z) = (x \wedge y) \vee (x \wedge z) \vee (y \wedge z) \quad \text{for all } x,y,z \in L$$

7.17. A lattice L is called modular if for all $x, y, z \in L$,

$$y \leq x \quad \text{implies} \quad x \wedge (y \vee z) = y \vee (x \wedge z) .$$

Of the two lattices in Exercise 7.4, the second one is not modular, and the first one is modular but not distributive. Every distributive lattice is modular.

7.18. A lattice L is called semimodular if for all $x, y \in L$,

$$x \text{ and } y \text{ cover } x \wedge y \quad \text{implies} \quad x \vee y \text{ covers } x \text{ and } y .$$

Every modular lattice is semimodular. If L is semimodular then for all $x, y \in L$,

$$x \text{ covers } x \wedge y \quad \text{implies} \quad x \vee y \text{ covers } y .$$

7.19. If A is a regular arithmetical convolution then N_A is a lattice if and only if $A = D$.

7.20. Let L_1 and L_2 be lattices and let $L = L_1 \times L_2$, the product of the posets L_1 and L_2 (see Exercise 7.11). Then L is a lattice. If L_1 and L_2 are modular (distributive) then L is modular (distributive).

7.21. Let L be a distributive lattice and let a, b, c, d be elements of L such that $c \wedge d \leq a \leq c$ and $c \wedge d \leq b \leq d$. The mapping $\eta : [a,c] \times [b,d] \to [a \vee b, c \vee d]$ defined by $\eta(\langle x,y \rangle) = x \vee y$ is one-one and onto. (Hint: consider the mapping $\eta' : [a \vee b, c \vee d] \to [a,c] \times [b,d]$ defined by $\eta'(z) = \langle c \wedge z, d \wedge z \rangle$.)

7.22. For the first lattice in Exercise 7.4, neither μ or $\zeta * \zeta$ is factorable.

7.23. Let L be a locally distributive local lattice and let $f \in F(L)$ have an inverse. Then f is factorable if and only if for all $a, c, d \in L$ belonging to an interval

$$f(a,c)f(a,d) = f(a,c \vee d) \quad \text{if} \quad c \wedge d = a$$

and

(a) $\qquad f(a,c) = f(a \vee d, c \vee d) \quad \text{if} \quad a \leq c \quad \text{and} \quad a \wedge d = c \wedge d .$

7.24. In Exercise 7.23, (a) can be replaced by

(b) $f(a \wedge b, a) = f(b, a \vee b)$ for all $a, b \in L$ belonging to an interval.

7.25. Continuing Exercises 7.23 and 7.24, if f is completely factorable and has an inverse, and if (b) holds for f, then f is factorable.

7.26. Let L be a local lattice for which (2) in the outline of the proof of Proposition 7.6 holds. Then neither μ or $\zeta * \zeta$ is factorable.

7.27. Let L be a local lattice. Then L is locally distributive if and only if the following is true:

if $f \in F(L)$ and if $g(x,y) = \sum_{x \leq z \leq y} f(z,y)$ for all $x, y \in L$,

then g is factorable if and only if f is factorable.

7.28. Let L be a local lattice and let $\tau \in F(L)$ be defined by

$$\tau(x,y) = \begin{cases} 1 & \text{if } y \text{ covers } x \\ 0 & \text{otherwise} . \end{cases}$$

Then τ is factorable if and only if every interval in L is totally ordered.

7.29. Let L be a local lattice. The function δ in $F(L)$ is factorable if and only if (1) in the outline of the proof of Proposition 7.6 does not hold (i.e., if and only if L is locally modular: see [C], p. 66).

7.30. Let L be a local lattice and let H be a function on L of the kind considered in the discussion of $\nu_{k,H}$. If h and k are non-negative integers let $\phi_{h,k,H} = \nu_{h,H}^{-1} * \nu_{k,H}$. Then $\phi_{h,k,H}^{-1} = \phi_{k,h,H}$ and if $x, y \in L$ and $x \leq y$,

$$\frac{H(y)^k}{H(x)^k} = \sum_{x\leq z\leq y} \phi_{h,k,H}(z,y) \frac{H(z)^h}{H(x)^h} .$$

7.31. If L is a locally distributive local lattice and if $f \in F(L)$ then $h = f * \mu$ is a factorable companion function for f .

7.32. If A is a regular arithmetical convolution and if $f(n) = \phi_A(n)/n$ for all n , then the associated function $\tilde{f} \in F(N_A)$ is constant on chains.

7.33. For all n and r

$$\sum_{\substack{d|r \\ (n,d)=1}} \frac{\phi(d)}{d} \mu(r/d) = \frac{\mu(r)}{m} , \quad m = \frac{r}{(n,r)} .$$

7.34. Theorem 2.5 is a corollary of Corollary 7.9.

7.35. Let A be a regular arithmetical convolution. Let f be a multiplicative function such that for every prime p and all $\alpha \geq 1$, $f(p^\alpha) = f(p^t)$ where $t = t_A(p^\alpha)$. Let h be a multiplicative function such that $h(p^\alpha) = f(p^\alpha) - 1$ for every primitive prime power p^α . Then for all n and r ,

$$\sum_{\substack{d\in A(r) \\ (n,d)_A=1}} f(d)\mu_A(r/d) = \mu_A(r)\mu_A(m)h(m) ,$$

where $m = r/(n,r)_A$.

7.36. An analogue of the Brauer-Rademacher theorem for regular arithmetical convolutions can be obtained from Exercise 7.35.

7.37. Let $\phi_{h,k,H}$ be the function defined in Exercise 7.30. If L is the lattice of Theorem 7.8, and if $x, y, t \in L$ and $x \leq y \quad t$, then

$$\sum_{\substack{x \leq z \leq y \\ z \wedge t = x}} \frac{\nu_{k,H}(x,z)}{\phi_{h,k,H}(x,z)} \mu(z,y) = \frac{\mu(x,y)\mu(y \wedge t,y)\nu_{h,H}(y \wedge t,y)}{\phi_{h,k,H}(y \wedge t,y)}$$

7.38. Under the hypotheses of Theorem 7.8, and if $f(x,y) \neq 0$ when y covers x , for all $x, y, t \in L$ with $x \leq y \wedge t$,

$$\sum_{\substack{x \leq z \leq y \\ z \wedge t = x}} h(x,z)\mu(x,z)^2 = \frac{f(x,y)}{f(x,y \wedge t)} .$$

7.39. Let N_0 be the set of nonnegative integers with the natural ordering. If $f, g \in F(N_0)$ and if $m \leq n$, then

$$(f * g)(m,n) = \sum_{j=m}^{n} f(j - m)g(n - j) .$$

f is factorable if and only if $f(n,n) = 1$ for all $n \in N_0$. If $H(n) = 2^n$ for all $n \in N_0$ then

$$\nu_{k,H}(m,n) = 2^{k(m-n)}$$

$$\nu^{-1}_{k,H}(m,n) = \begin{cases} 1 & \text{if } m = n \\ -2^k & \text{if } m = n-1 \\ 0 & \text{otherwise} \end{cases}$$

$$\sigma_{k,H}(m,n) = 1 + 2^k + 2^{2k} + \ldots + 2^{k(n-m)} ,$$

$$\sigma^{-1}_{k,H}(m,n) = \begin{cases} 1 & \text{if } m = n \\ -2^k - 1 & \text{if } m = n-1 \\ 2^{2k} & \text{if } m = n-2 \\ 0 & \text{otherwise} \end{cases}$$

$$\phi_{h,k,H}(m,n) = \begin{cases} 1 & \text{if } m = n \\ 2^{k(m-n)} - 2^{k(m-n-1)+h} & \text{if } m < n . \end{cases}$$

7.40. If m and n are nonnegative integers let $\binom{n}{m}$ be the binomial coefficient, i.e.,

$$\binom{n}{m} = \begin{cases} \dfrac{n!}{m!(n-m)!} & \text{if } m \leq n \\ 0 & \text{if } m > n . \end{cases}$$

Let p be a prime and write $m \leq_p n$ if $p \nmid \binom{n}{m}$. If

$$m = a_0 + a_1 p + \ldots \qquad , \quad 0 \leq a_i \leq p - 1 ,$$

$$n = b_0 + b_1 p + \ldots \qquad , \quad 0 \leq b_i \leq p - 1 ,$$

then $m \leq_p n$ if and only if $a_i \leq b_i$ for $i = 0, 1, \ldots$. (Hint: show that

$$\binom{n}{m} \equiv \binom{b_0}{a_0}\binom{b_1}{a_1} \cdots \pmod{p}.)$$

Thus, $\leq_p$ is a partial ordering on the set of nonnegative integers. The resulting poset, L , is a locally finite lattice:

$$m \vee n = c_0 + c_1 p + \ldots ,$$

$$m \wedge n = d_0 + d_1 p + \ldots ,$$

where $c_i = \max(a_i, b_i)$ and $d_i = \min(a_i, b_i)$. L is distributive.

7.41. Continuing the preceding exercise, if $m \leq_p n$ then

$$\mu(m,n) = \begin{cases} (-1)^s & \text{if } n - m \text{ is the sum of } s \text{ distinct powers of } p \\ 0 & \text{otherwise.} \end{cases}$$

7.42. Let B be a basic sequence (see Exercise 4.41). The relation $\leq$ on the set of positive integers defined by

$$m \leq n \quad \text{if } m|n \text{ and } \langle m, n/m \rangle \in B$$

is a partial ordering, and the resulting locally finite poset will be denoted by N_B . If f is an arithmetical function and if $\tilde{f} \in F(N_B)$ is defined by

$$\tilde{f}(m,n) = \begin{cases} f(n/m) & \text{if } m \leq n \\ 0 & \text{otherwise ,} \end{cases}$$

then for arithmetical functions f and g and for all n,

$$(f *_B g)(n) = \sum_{1 \le_B d \le_B n} \tilde{f}(1,d)\tilde{g}(d,n) .$$

7.43. This exercise and the ones which follow contain results from combinatorial theory. Let S be a nonempty finite set and let w be a real-valued function on S: $w(x)$ is called the <u>weight</u> of x. Let $A_1,\ldots,A_n$ be subsets of S. Let P be the poset consisting of all subsets of $\{1,\ldots,n\}$, ordered by set inclusion. If $I \in P$ let $f(I)$ be the sum of the weights of the elements x of S such that $x \in A_i$ if and only if $i \notin I$, and let

$$g(I) = \sum_{J \subseteq I} f(J).$$

If $0 \le m \le n$ then

$$\sum_{\#I=n-m} f(I) = \sum_{j=0}^{n-m} (-1)^{n-m-j} \binom{n-j}{n-m-j} \sum_{\#J=j} g(J).$$

(Hint: use Exercises 7.6 and 7.10.)

7.44. Continuing Exercise 7.43, if $\{i_1,\ldots,i_k\} \subseteq \{1,\ldots,n\}$, let $N(i_1,\ldots,i_k)$ be the sum of the weights of the elements in $A_{i_1} \cap \ldots \cap A_{i_k}$ and let $N_k = \sum N(i_1,\ldots,i_k)$, where the sum is over all k-element subsets of $\{1,\ldots,n\}$. Let $N(m)$ be the sum of the weights of those elements of S which are in exactly m of the sets $A_1,\ldots,A_n$. Then

$$N(m) = \sum_{k=m}^{n} (-1)^{k-m} \binom{k}{m} N_k.$$

if $w(x) = 1$ for all $x \in S$ and $m = 0$, this is the Inclusion-Exclusion Principle. (Hint:

$$N_k = \sum_{\#J=n-k} g(J) \quad \text{and} \quad N(m) = \sum_{\#I=n-m} f(I).)$$

7.45. A permutation $j_1, j_2, \ldots, j_n$ of $1,2,\ldots,n$ is called a <u>derangement</u> if $j_i \neq i$ for $i = 1,\ldots,n$. If D_n is the number of derangements then

$$D_n = n! \sum_{k=0}^{m} (-1)^k \frac{1}{k!} .$$

(Hint: consider the set S of all permutations of $1,2,\ldots,n$ and let A_i be the set of permutations $j_1, j_2, \ldots, j_n$ such that $j_i = i$.)

7.46. If L is a finite lattice with least element 0 and if $a,b \in L$ and $b > 0$, then

$$\sum_{x \vee b=a} \mu(0,x) = 0.$$

If L has greatest element 1 and if $c,d \in L$ and $d < 1$, then

$$\sum_{x \wedge d=c} \mu(x,1) = 0.$$

If L is an arbitrary lattice and if a and b are distinct elements of L such that the interval $[a,b]$ is finite, then for all $c \in [a,b]$,

$$\mu(a,b) = - \sum_{\substack{a<x\leq b \\ x \wedge c=a}} \mu(x,b).$$

7.47. This exercise requires some knowledge of finite dimensional vector spaces. Let q be a power of a prime and let V_n be an n-dimensional vector space over the field with q elements. Let $L(V_n)$ be the lattice of subspaces of V_n: the lattice operations are

$$X \wedge Y = X \cap Y \quad \text{and} \quad X \vee Y = X + Y.$$

If $X, Y \in L(V_n)$, $X \subseteq Y$, then the interval $[X, Y]$ and $L(Y/X)$ are isomorphic as lattices. Thus, to determine the Möbius function of $L(V_n)$ for all n, it is sufficient to determine $\mu(0, V_n)$ for all n. (See Exercise 7.9.) In fact,

$$\mu(0, V_n) = (-1)^n q^{\frac{1}{2} n(n-1)}.$$

(Hint: use induction on n. The formula holds for $n = 1$. Assume $n > 1$ and make the proper induction assumption. Let Y be an $(n-1)$-dimensional subspace of V_n. If X is a non-zero subspace of V_n then $X \cap Y = 0$ if and only if X is 1-dimensional and $X \not\subseteq Y$. Use Exercise 7.46 and count the X's.)

7.48. This exercise requires some knowledge of finite group theory. Let G be a finite group and let p be an odd prime. If $f_k(G)$ is the number of pairs of subgroups of G of order p^k that generate G, then $f_k(G) \equiv 0 \pmod{p}$. Of course, it can happen that $f_k(G) = 0$. (Hint: the set of subgroups of G, ordered by inclusion, is a lattice. Determine $\sum_{K \subseteq H} f_k(K)$ for subgroups H of G, and use Exercise 7.6.)

Notes on Chapter 7

The earliest papers on Möbius functions and inversion theorems in settings other than that of number theory were concerned with lattices of subgroups of a group. L. Weisner [35a], [35b] and P. Hall [34] were interested in prime-power groups and S. Delsarte [48] and E. Cohen [61j], [62c] in Abelian groups. M. Ward [39] worked on a more abstract level. He defined the convolution of incidence functions of a locally finite lattice, and pointed out that the incidence functions form a ring with respect to addition and convolution. See also the paper by R. Wiegandt [59].

In 1964, G.-C. Rota [64] laid the foundations of the theory of the Mobius functions in combinatorial mathematics. His paper is filled with historical comments and examples, and it has an extensive bibliography. It triggered a burst of activity in the area, and today there is a discussion of the Möbius function in virtually every modern text in combinatorial mathematics. A number of applications of the Möbius function are contained in the paper by E. A. Bender and J. R. Goldman [75].

The definition of a factorable function was given by D. Smith [67], and he proved Propositions 7.4 and 7.5. He proved these propositions for distributive lattices, but in another paper [69a] he noted that they hold for locally distributive local lattices.

M. Ward [39] called an incidence function of a locally finite lattice L factorable if

$$f(a,b)f(c,d) = f(a \wedge c,\ b \wedge d)f(a \vee c,\ b \vee d)$$

for all $a,b,c,d \in L$ with $a \leq b$ and $c \leq d$. He showed that if $\zeta * \zeta$ is factorable then L is distributive, but he did not give conditions for the convolution of factorable functions to be factorable. With his weaker definition of factorable incidence function, D. Smith [67], [69a] proved Theorem 7.7, and also the results of Exercises 7.27-7.29. In another paper [69b] he obtained the facts concerning factorable and completely factorable functions in Exercises 7.23-7.25: the result on completely factorable functions in Exercise 7.13 was observed by P. J. McCarthy [70]. Note that for the incidence functions of the lattice N_D, the two definitions of factorable function are equivalent: this is the gist of Exercises 1.9 and 1.10.

The functions $\nu_{k,H}$, $\sigma_{k,H}$ and $\phi_{h,k,H}$ (in Exercise 7.30) were introduced by D. Smith [67], and he worked out the examples in Exercises 7.39-7.41 (see also papers by L. Carlitz [66a], [67]). In another paper [69b] he proved Theorem 7.8, the results in Exercises 7.37 and 7.38 and other, similar results which are abstract versions of arithmetical identities. There is a parallel development of the theory of incidence functions of locally finite posets in papers of H. Scheid [68a], [68b], [69b], [70], [71]. He obtained also lattice-theoretic analogues of such well-known identities as the Brauer-Rademacher identity and, furthermore, he defined a Ramanujan-type sum and developed a theory of even functions.

The inversion theorems of Exercises 7.6 and 7.7, as well as the results in Exercises 7.46 and 7.48 are due to L. Weisner [35a], [35b]. In the second paper he determined the Möbius function of the lattice of all subgroups of a group whose orders are powers of a prime p.

Finally, in addition to the expository papers by G.-C. Rota [64] and by E. A. Bender and J. R. Goldman [75], which emphasize combinatorial applications, there is one by D. Smith [72] on his work and that of H. Scheid, and which has an arithmetical emphasis.

References

[Ah] L. V. Ahlfors, Complex Analysis, Second Edition, McGraw-Hill Book Co., New York, etc., 1966.

[A] T. M. Apostol, Introduction to Analytic Number Theory, Springer-Verlag, New York, Heidelberg, Berlin, 1976.

[A'] T. M. Apostol, Modular Functions and Dirichlet Series in Number Theory, Springer-Verlag, New York, Heidelberg, Berlin, 1976.

[Ay] R. Ayoub, An Introduction to the Analytic Theory of Numbers, American Mathematical Society, Providence, 1963.

[C] P. M. Cohn, Universal Algebra, Harper & Row, New York, Evanston, London, 1965.

[D] L. E. Dickson, History of the Theory of Numbers, Volume I, Carnegie Institution of Washington, 1919; reprinted, Chelsea Publishing Co., New York, 1952.

[H] G. H. Hardy, Ramanujan, Cambridge University Press, Cambridge, 1940; reprinted, Chelsea Publishing Co., New York, 1959.

[HW] G. H. Hardy and E. M. Wright, An Introduction to the Theory of Numbers, Fifth Edition, Oxford University Press, Oxford, 1979.

[L] E. Landau, Handbuch der Lehre von der Verteilung der Primzahl, B. G. Teubner, Leipzig, Berlin, 1909; reprinted, Chelsea Publishing Co., New York, 1974.

[NZ] I. Niven and H. S. Zuckerman, An Introduction to the Theory of Numbers, John Wiley & Sons, New York, London, Sydney, 1960.

[R] E. D.Rainville, Special Functions, The Macmillan Co., New York, 1960; reprinted, Chelsea Publishing Co., New York, 1971.

Bibliography

H. L. Adler
[58] A generalization of the Euler ϕ-function, Amer. Math. Monthly 65(1958) 690-692

K. Alladi
[77] On arithmetic functions and divisors of higher order, J. Austral. Math. Soc. 23A(1977) 9-27

D. R. Anderson and T. M. Apostol
[53] The evaluation of Ramanujan's sum and generalizations, Duke Math. J. 20(1953) 211-216

T. M. Apostol
[65] A characteristic property of the Möbius function, Amer. Math. Monthly 72(1965) 279-282
[70] Möbius functions of order k, Pacific J. Math. 32(1970) 279-282
[71] Some properties of the completely multiplicative arithmetic functions, Amer. Math. Monthly 78(1971) 266-271
[72] Arithmetical properties of generalized Ramanujan sums, Pacific J. Math. 41(1972) 281-293
[73] Identities for series of the type $\Sigma f(n)\mu(n)n^{-s}$, Proc. Amer. Math. Soc. 40(1973) 341-345
[75a] Note on series of type $\Sigma f(n)\mu(n)n^{-s}$, Nordisk Mat. Tidskr. 23(1975) 49-50
[75b] A note on periodic completely multiplicative arithmetical functions, Amer. Math. Monthly 83(1975) 39-40
[77] Arithmetical subseries of series of multiplicative terms, Bull. Soc. Math. Grece 18(1977) 106-120

T. M. Apostol and H. S. Zuckerman
[64] On the functional equation $F(mn)F((m,n)) = F(m)F(n)f((m,n))$, Pacific J. Math. 14(1964) 377-384

A. Baloq
[81] On a conjecture of A. Ivić and W. Schwarz, Publ. Inst. Math. (Beograd) (N.S.) 30(44)(1981) 11-15

E. T. Bell
[15] Arithmetical theory of certain numerical functions, Univ. Wash. Publ. Math. Phys. Sci. Vol. I, No. 1, 1915
[18] Numerical functions of [x], Ann. of Math. 19(1918) 210-216
[20] On a certain inversion in the theory of numbers, Tohoku Math. J. 17(1920) 221-231
[21] Proof of an arithmetic theorem due to Liouville, Bull. Amer. Math. Soc. 27(1921) 273-275
[22] Extensions of Dirichlet multiplication and Dedekind inversion, Bull. Amer. Math. Soc. 29(1922) 111-122

[24] Sur l'inversion des produits arithmétiques, Enseign. Math. 23(1924) 305-308
[28] Outline of a theory of arithmetical functions in the algebraic aspects, J. Indian Math. Soc. 17(1928) 249-260
[31a] Note on functions of r-th divisors, Amer. J. Math. 53(1931) 56-60
[31b] Functional equations for totients, Bull. Amer. Math. Soc. 37(1931) 85-90
[31c] Factorability of numerical functions, Bull. Amer. Math. Soc. 37(1931) 251-253; Addendum, ibid. 630
[34] An algebra of numerical compositions, J. Indian Math. Soc. 29(1934) 129-238
[36] Note on an inversion formula, Amer. Math. Monthly 43(1936) 464-465
[37] General theorems on numerical functions, J. Math. Pures Appl. 16(1937) 151-154

R. Bellman
[50] Ramanujan sums and the average value of arithmetic functions, Duke Math. J. 17(1950) 159-168

E. A. Bender and J. R. Goldman
[75] On the applications of Möbius inversion to combinatorial analysis, Amer. Math. Monthly 82(1975) 789-803

S. J. Benkoski
[76] The probability that k positive integers are relatively prime, J. Number Theory 8(1976) 218-223

M. G. Beumer
[62] The arithmetical function $\tau_k(N)$, Amer. Math. Monthly 69(1962) 777-781

F. van der Blij
[50] The function $\tau(n)$ of S. Ramanujan (an expository lecture), Math. Student 18(1950) 83-99

A. Brauer
[26] Lösungen der Aufgaben 30-32, Jahr. Deutsch. Math.-Verein., 35(1926), 2. Abteilung, 92-96

E. Brinitzer
[75] Über (k,r)-Zahlen, Monatsh. Math. 80(1975) 31-35

R. G. Buschman
[70] Identities involving Golubev's generalization of the μ-function, Portugal. Math. 29(1970) 145-149

L. Carlitz

[37] An arithmetic function, Bull. Amer. Math. Soc. 43(1937) 271-276

[52a] Independence of arithmetic functions, Duke Math. J. 19(1952) 65-70

[52b] Note on an arithmetic function, Amer. Math. Monthly 59(1952) 386-387

[64] Rings of arithmetic functions, Pacific J. Math. 14(1964) 1165-1171

[66a] Arithmetic functions in an unusual setting, Amer. Math. Monthly 73(1966) 582-590

[66b] A note on the composition of arithmetic functions, Duke Math. J. 33(1966) 629-632

[67] Arithmetic functions in an unusual setting II, Duke Math. J. 34(1967) 757-759

[69] Sums of arithmetic functions, Collect. Math. 20(1969) 107-126

[71] Problem E2268, Amer. Math. Monthly 78(1971) 1140

R. D. Carmichael

[30] On expansions of arithmetical functions, Proc. Nat. Acad. Sci. U.S.A. 16(1930) 613-616

[32] Expansions of arithmetical functions in infinite series, Proc. London Math. Soc. 34(1932) 1-26

T. B. Carroll

[74] A characterization of completely multiplicative arithmetic functions, Amer. Math. Monthly 81(1974) 993-995

T. B. Carroll and A. A. Gioia

[75] On a subgroup of the group of multiplicative arithmetic functions, J. Austral. Math. Soc. Ser. A 20(1975) 348-358

E. Cashwell and C. Everett

[59] The ring of number-theoretic functions, Pacific J. Math. 9(1959) 975-985

J. Chidambaraswamy

[67a] Sum functions of unitary and semi-unitary divisors, J. Indian Math. Soc. 31(1967) 117-126

[67b] On the functional equation F(mn)F((m,n)) = F(m)F(n)f((m,n)), Portugal. Math. 26(1967) 101-107

[70] The K-unitary convolution of certain arithmetical functions, Publ. Math. Debrecen 17(1970) 67-74

[74] Totients with respect to a polynomial, Indian J. Pure Appl. Math. 5(1974) 601-608

[75] A remark on a couple of papers of D. Suryanarayana, Math. Student 43(1975) 39-41

[76a] Generalized Dedekind ψ-function with respect to a polynomial. I, J. Indian Math. Soc. 18(1976) 23-34

[76b] Generalized Dedekind ψ-function with respect to a polynomial. II, Pacific J. Math. 65(1976) 19-27

[77] Series involving the reciprocals of generalized totients, J. Nat. Sci. Math. 17(1977) 11-26

[79a] Totients and unitary totients with respect to a set of polynomials, Indian J. Pure Appl. Math. 10(1979) 287-302

[79b] Generalized Ramanujan's sum, Periodica Math. Hungarica 19(1979) 71-87

J. Chidambaraswamy and Sitaramachandra Rao

[83] On arithmetical functions associated with higher order divisors, J. Okayama Univ. 26(1983) 185-194

S. D. Chowla

[28] On some identities involving zeta-functions, J. Indian Math. Soc. 17(1928) 153-163

[29] Some identities in the theory of numbers, J. Indian Math. Soc., Notes & Questions 18(1929) 87-88

J. Christopher

[56] The asymptotic density of some k-dimensional sets, Amer. Math. Monthly 63(1956) 399-401

E. Cohen

[49] An extension of Ramanujan's sum, Duke Math. J. 16(1949) 85-90

[52] Rings of arithmetic functions, Duke Math. J. 19(1952) 115-129

[53] Congurence representations in algebraic number fields, Trans. Amer. Math. Soc. 75(1953) 444-470

[54] Rings of arithmetic functions. II. The number of solutions of quadratic congruences, Duke Math. J. 21(1954)

[55a] An extension of Ramanujan's sum. II. Additive properties, Duke Math. J. 22(1955) 543-550

[55b] A class of arithmetical functions, Proc. Nat. Acad. Sci. U.S.A. 41(1955) 939-944

[56a] An extension of Ramanujan's sum. III. Connections with totient functions, Duke Math. J. 23(1956) 623-630

[56b] Some totient functions, Duke Math. J. 25(1956) 515-522

[58a] Generalizations of the Euler ϕ-function, Scripta Math. 23(1958) 157-161

[58b] Representations of even functions (mod r). I. Arithmetical identities, Duke Math. J. 25(1958) 401-421

[59a] Trigonometric sums in elementary number theory, Amer. Math. Monthly 66(1959) 105-117

[59b] Representations of even functions (mod r). II. Cauchy products, Duke Math. J. 26(1959) 165-182

[59c] An arithmetical inversion principle, Bull. Amer. Math. Soc. 65(1959) 335-336

[59d] Representations of even functions (mod r). III. Special topics, Duke Math. J. 26(1959) 491-500

[59e] A class of residue systems (mod r) and related arithmetical functions. I. A generalization of Möbius inversion, Pacific J. Math. 9(1959) 13-23

[59f] A class of residue systems (mod r) and related arithmetical functions. II. Higher dimensional analogues, Pacific J. Math. 9(1959) 667-679

[59g] Arithmetical functions associated with arbitrary sets of integers, Acta Arith. 5(1959) 407-415

[60a] The Brauer-Rademacher identity, Amer. Math. Monthly 67(1960) 30-33

[60b] Arithmetical functions of a greatest common divisor. I, Proc. Amer. Math. Soc. 11(1960) 164-171

[60c] Arithmetical functions associated with the unitary divisors of an integer, Math. Z. 74(1960) 66-80

[60d] Arithmetical inversion formulas, Canad. J. Math. 12(1960) 399-409

[60e] A class of arithmetical functions in several variables with applications to congruences, Trans. Amer. Math. Soc. 96(1960) 335-381

[60f] The elementary arithmetical functions, Scripta Math. 25(1960) 221-227

[60g] The average order of certain types of elementary arithmetical functions: generalized k-free numbers and totient points, Monatsh. Math. 64(1960) 251-262

[60h] The number of unitary divisors of an integer, Amer. Math. Monthly 67(1960) 879-880

[60i] Arithmetical functions of finite abelian groups, Math. Ann. 142(196)]65-]82

[60j] Nagell's totient function, Math. Scand. 8(1960) 55-58

[60k] A trigonometric sum, Math. Student 28(1960) 29-32

[61a] Unitary products of arithmetical functions, Acta Arith. 7(1961) 29-38

[61b] Arithmetical notes. I. On a theorem of van der Corput, Proc. Amer. Math. Soc. 12(1961) 214-217

[61c] Fourier expansions of arithmetical functions, Bull. Amer. Math. Soc. 12(1961) 214-217

[61d] Series representations of certain types of arithmetical functions, Osaka Math. J. 13(1961) 209-216

[61e] An elementary method in the asymptotic theory of numbers, Duke Math. J. 28(1961) 183-192

[61f] Arithmetical functions of a greatest common divisor. III. Cesàro's divisor problem, Proc. Glasgow Math. Assoc. 5(1961) 67-75

[61g] Arithmetical note. V. A divisibility property of the divisor function, Amer. J. Math. 83(1961) 693-697

[61h] Some sets of integers related to the k-free integers, Acta Sci. Math. (Szeged) 22(1961) 223-233

[61i] A property of Dedekind's ψ-function, Proc. Amer. Math. Soc. 12(1961) 996

[61j] On the inversion of even functions of finite abelian groups, J. Reine Angew. Math. 207(1961) 192-202

[61k] Unitary functions (mod r), Duke Math. J. 28(1961) 475-485

[62a] An identity related to the Dedekind-von Sterneck function, Amer. Math. Monthly 69(1962) 213-215

[62b] Arithmetical notes. VIII. An asymptotic formula of Renyi, Proc. Amer. Math. Soc. 13(1962) 536-539

[62c] Almost even functions of finite abelian groups, Acta. Arith. 7(1962) 311-323

[62d] Unitary funcions (mod r). II. Publ. Math. Debrecen 9(1962) 94-104
[62e] Averages of completely even arithmetical functions (over certain types of plane regions), Ann. Mat. Pura Appl. 59(1962) 165-177
[62f] Arithmetical functions of a greatest common divisor. II. An alternative approach, Boll. Un. Mat. Ital. 17(1962) 349-356
[63a] On the distribution of certain sequences of integers, Amer. Math. Monthly 70(1963) 516-521
[63b] An elementary estimate for the k-free integers, Bull. Amer. Math. Soc. 69(1963) 762-765
[63c] Arithmetical notes. II. An estimate of Erdös and Szekeres, Scripta Math. 26(1963) 353-356
[63d] Some analogues of certain arithmetical functions, Riv. Mat. Univ. Parma 4(1963) 115-125
[64a] Arithmetical notes. X. A class of totients, Proc. Amer. Math. Soc. 15(1964) 534-539
[64b] A generalization of Axer's theorem and some of its applications, Math. Nachr. 17(1964) 163-177
[64c] Arithmetical notes, XI. Some divisor identities, Enseignement Math. 10(1964) 248-254
[64d] Some asymptotic formulas in the theory of numbers, Trans. Amer. Math. Soc. 112(1964) 214-227
[64e] Remark on a set of integers, Acta Sci. Math. (Szeged) 25(1964) 179-180
[64f] Arithmetical notes. VII. Some classes of even functions (mod r), Collect. Math. 16(1964) 81-87
[68] A generalized Euler ϕ-function, Math. Mag. 41(1968) 276-279
[69] On the mean parity of arithmetical functions, Duke Math. J. 36(1969) 659-668

E. Cohen and K. J. Davis
[70] Elementary estimates for certain types of integers, Acta Sci. Math. (Szeged) 31(1970) 363-371

E. Cohen and R. L. Robinson
[63] On the distribution of k-free integers in residue classes, Acta Arith. 8(1963) 283-293; errata, ibid. 10(1964) 443; correction, ibid. 16(1970) 439

P. Comment
[57] Sur l'equation fonctionelle F(nm)F((n,m)) = F(n)F(m)f((n,m)), Full. Res. Council Israel Sect. F 17F(1957/58) 14-20

J. G. van der Corput
[39] Sur quelques fonctions arithmétiques élémentaires, Proc. Kon. Nederl. Akad. Wetensch. 42(1939) 859-866

M. M. Crum
[40] On some Dirichlet series, J. London Math. Soc. 15(1940) 10-15

K. J. Davis
[82] A generalization of the Dirichlet product, Fibonacci Quart. 20(1982) 41-44

T. M. K. Davison
[66] On arithmetic convolutions, Canad. Math. Bull. 9(1966) 287-296

D. E. Daykin
[64] Generalized Möbius inversion formulae, Quart. J. Math. 15(1964) 349-354
[65] An arithmetic congruence, Amer. Math. Monthly 72(1965) 291-292
[72] A generalization of the Möbius inversion formula, Simon Stevin 46(1972) 141-146

J.-M. DeKoninck and A. Mercier
[77] Remarque sue le article "Identities for series of the type $\Sigma f(n)\mu(n)n^{-s}$" (Proc. Amer. Math. Soc. 40(1973) 341-345) de T. M. Apostol, Canad. Math. Bull. 20(1977) 77-88

H. Delange
[76] On Ramanujan expansions of certain arithmetical functions, Acta Arith. 31(1976) 259-270

S. Delsarte
[48] Fonctions de Möbius sur les groupes abeliens finis, Ann. of Math. 49(1948) 600-609

J. D. Dixon
[60] A finite analogue of the Goldbach problem, Canad. Math. Bull. 3(1960) 121-126

G. S. Donavan and D. Rearick
[66] On Ramanujan's sum, Norske Vid. Selsk. Forh. (Trondheim) 39(1966) 1-2

P. Erdös
[48] Some asymptotic formulas in number theory, J. Indian Math. Soc. 12(1948) 75-78

P. Erdös and G. G. Lorentz
[58] On the probability that n and g(n) are relatively prime, Acta Arith. 5(1958) 35-44

F. Eugeni
[73] Numeri primitivi e indicator generalizzati, Rend. Math. (6) 6(1973) 97-130

F. Eugeni and B. Rizzi
[79a] Una classe di funzioni arithmetiche periodiche, La Ricerca 30, No. 2, (1979) 3-14

[79b] Una estensione di due identita' di Anderson-Apostol e Landau La Ricerca 30, No. 3, (1979) 11-18
[79c] An incidence algebra on rational numbers, Rend. Math. (6) 12(1979) 557-576
[80a] Somma generalizzata di Ramanujan per l'indicatore di Stevens, La Ricerca 31, No. 3, (1980) 3-10
[80b] On certain solutions of Cohen functional equation relating the Brauer-Rademacher identity, Boll. Un. Mat. Ital. 17B(1980) 969-978

J. A. Ewell
[84] A formula for Ramanujan's tau function, Proc. Amer. Math. Soc. 91(1984) 37-40

G. W. Fang
[81] Some identities of Dirichlet series relating with the Möbius function, Chinese J. Math. 9(1981) 79-85

I. P. Fatino
[75] Generalized convolution ring of arithmetical functions, Pacific J. Math. 61(1975) 103-116

M. Ferrero
[80] On generalized convolution rings of arithmetic functions, Tsubuka J. Math. 4(1980) 161-176

O. M. Fomenko
[61a] Quelques remarques sur la fonction de Jordan, Mathesis 69(1961) 69(1961) 287-291

O. M. Fomenko and V. A. Golubev
[61] On the functions $\phi_2(n)$, $\mu_2(n)$, $\zeta_2(n)$, Ann. Polon. Math. 11(1961) 13-17 (Russian)

M. L. Fredman
[70] Arithmetical convolution products, Duke Math. J. 37(1970) 231-242

E. Gagliardo
[53] Le funzioni simmetriche semplici delle radici n-esime primitive dell'unità, Boll. Un. Mat. Ital. 8(1953) 269-273

M. D. Gessley
[67] A generalized arithmetic convolution, Amer. Math. Monthly 74(1967) 1216-1217

A. A. Gioia
[62] On an identity for multiplicative functions, Amer. Math. Monthly 69(1962) 988-991
[65] The K-product of arithmetic functions, Canad. J. Math. 17(1965) 970-976

A. A. Gioia and D. L. Goldsmith
[71] Convolutions of arithmetic functions over cohesive basic sequences, Pacific J. Math. 38(1971) 391-399

A. A. Gioia and M. V. Subbarao
[62] Generlized Dirichlet products of arithmetic functions (abstract), Notices Amer. Math. Soc. 9(1962) 305
[66] Generating functions for a class of arithmetic functions, Canad. Math. Bull. 9(1966) 427-431

A. A. Gioia and A. M. Vaidya
[66] The number of square-free divisors of an integer, Duke Math. J. 33(1966) 797-799

R. R. Goldberg and R. S. Varga
[56] Möbius inversion and Fourier transforms, Duke Math. J. 23(1956) 553-559

D. L. Goldsmith
[68] On the multiplicative properties of arithmetic functions, Pacific J. Math. 27(1968) 283-304
[69] A generalization of some identities of Ramanujan, Rend. Mat. (6) 2(1969) 472-479
[70a] A note on sequences of almost-multiplicative arithmetic functions, Rend. Mat. (6) 3(1970) 167-170
[70b] On the structure of certain basic sequences associated with an arithmetic function, Proc. Edinburgh Math. Soc. (2) 17(1970/71) 305-310
[71] A generalized convolution for arithmetic functions, Duke Math. J. 38(1971) 279-283
[72] On the density of certain cohesive basic sequences, Pacific J. Math. 42(1972) 323-327
[74] Minimal cohesive basic sets, Proc. Edinburgh Math. Soc.(2) 19(1974) 73-76

D. L. Goldsmith and A. A. Gioia
[72] On a question of Erdös concerning cohesive basic sequences, Proc. Amer. Math. Soc. 34(1972) 356-358

F. Götze
[68] Über Teilerfunktionen in Dirichletschen Reihen, Arch. Math. 19(1968) 627-634

A. Grytczuk
[81] An identity involving Ramanujan's sum, Elem. Math. 36(1981) 16-17

E. E. Guerin
[78] Matrices and convolutions of arithmetic functions, Fibonacci Quart. 16(1978) 327-334

B. Gyires

[49] Über die Faktorisation im Restklassenring mod m, Publ. Math. Debrecen 1(1949) 51-55

P. Hall

[34] A contribution to the theory of groups of prime power order, Proc. London Math. Soc. 36(1934) 24-80

R. R. Hall

[70] On the probability that n and f(n) are relatively prime, Acta Arith. 17(1970) 169-183; correction, ibid. 19(1971) 203-204

[71] On the probability that n and f(n) are relatively prime. II. Acta Arith. 19(1971) 175-184

[72] On the probability that n and f(n) are relatively prime. III. Acta Arith. 20(1972) 267-289

L. van Hamme

[71] Sur une généralisation de l'indicateur d'Euler, Acad. Roy. Belg. Bull. Cl. Sci. (5) 57(1971) 805-817

R. T. Hansen

[80] Arithmetic inversion formulas, J. Natur. Sci. Math. 20(1980) 141-150

R. T. Hansen and L. G. Swanson

[80] Vinogradov's Möbius inversion theorem, Nieuw Arch. Wisk. 28(1980) 1-11

J. Hanumanthachari

[70a] Some generalized unitary functions and their applications to linear congruences, J. Indian Math. Soc. 34(1970) 99-108

[70b] Certain generalizations of Nagell's totient function and Ramanujan's sum, Math. Student 38(1970) 183-187: correction and addition, ibid. 42(1974) 121-124

[72] A generalization of Dedekind's Ψ-function, Math. Student 40A(1972) 1-4

[77] On an arithmetic convolution, Canad. Math. Bull. 20(1977) 301-305

J. Hanumanthachari and V. V. Subrahmanyasastri

[74] Certain totient functions - allied Ramanujan sums and applications to certain linear congruences, Math. Student 42(1974) 214-222

[78] On some arithmetical identities, Math. Student 46(1978) 60-70

G. H. Hardy

[21] Note on Ramanujan's trigonometrical function $c_q(n)$, and certain series of arithmetical functions, Proc. Camb. Philos. Soc. 20(1921) 263-271

E. K. Haviland

[44] An analogue of Euler's ϕ-function, Duke Math. J. 11(1944) 869-872

M. Hayashi

[80] On some Dirichlet series generated by an arithmetic function, Mem. Osaka Inst. Tech. Ser. A. 25(1980) 1-5

E. Hille

[37] The inversion problem of Möbius, Duke Math. J. 3(1937) 549-569

E. Hille and O. Szász

[36a] On the completeness of Lambert functions, Bull. Amer. Math. Soc. 42(1936) 411-418

[36b] On the completeness of Lambert functions. II. Ann. of Math. 37(1936) 801-815

O. Hölder

[36] Zur Theorie der Kreisteilungsgleichung $K_m(x) = 0$, Prace Mat.-Fiz. 43(1936) 13-23

E. M. Horadam

[61] Arithmetical functions of generalized primes, Amer. Math. Monthly 68(1961) 626-629

[62] Arithmetical functions associated with the unitary divisors of a generalized integer, Amer. Math. Monthly 69(1962) 196-199

[63a] The order of arithmetical functions of generalized integers, Amer. Math. Monthly 70(1963) 506-512

[63b] The Euler ϕ function for generalized integers, Proc. Amer. Math. Soc. 14(1963) 754-762

[63c] A calculus of convolutions for generalized integers, Proc. Nederl. Akad. Wetensch. 66(1963) 695-698

[64a] Ramanujan's sum for generalized integers, Duke Math. J. 31(1964); Addendum, ibid. 33(1966) 705-707

[64b] The number of unitary divisors of a generalized integer, Amer. Math. Monthly 71(1964) 893-895

[65] Unitary divisor functions of a generalized integer, Portugal Math. 24(1965) 131-143

[66] Exponential functions for arithmetical semigroups, J. Reine Angew. Math. 222(1966) 14-19

[67] A sum of a certain divisor function for arithmetical semigroups, Pacific J. Math. 22(1967) 407-412

[68] Ramanujan's sum and its applications to the enumerative functions of certain sets of elements of an arithmetical semigroup, J. Math. Sci. 3(1968) 47-70

[71] An extension of Daykin's generalized Möbius function to unitary divisors, J. Reine Angew. Math. 246(1971) 117-125

A. Ivić

[76] On a class of arithmetical functions connected with multiplicative functions, Publ. Inst. Math. (Beograd) (N.S.) 20(34)(1976) 131-144

[77] A property of Ramanujan's sums concerning totally multiplicative functions, Univ. Beograd Publ. Elektrotehn. Fak. Ser. Mat. Fiz. No. 577-598 (1977) 74-78

A. Ivić and W. Schwarz

[80] Remarks on some number-theoretical functional equations, Aequationes Math. 20(1980) 80-89

H. Jager

[61] The unitary analogues of some identities for certain arithmetical functions, Nederl. Akad. Wetensch. Proc. Ser. A 64(1961) 508-515

K. R. Johnson

[82a] A reciprocity law for Ramanujan sums, Pacific J. Math. 98(1982) 99-105

[82b] Unitary analogues of generalized Ramanujan sums, Pacific J. Math. 103(1982) 429-436

[83] A result for the "other" variable of Ramanujan's sum, Elem. Math. 38(1983) 122-124

G. S. Kazandzidis

[63] Algebra of subsets and Möbius pairs, Math. Ann. 152(1963) 208-225

P. Kemp

[75] A note on a generalized Möbius function, J. Natur. Sci. Math. 15(1975) 55-57

[76a] Properties of multiplicative functions, J. Natur. Sci. Math. 16(1976) 51-56

[76b] Extensions of the ring of number theoretic functions, J. Natur. Sci. Math. 16(1976) 57-60

[77] Nth roots of arithmetical functions, J. Natur. Sci. Math. 17(1977) 27-30

P. Kesava Menon

[42a] Multiplicative functions which are functions of the g.c.d. and l.c.m. of their arguments, J. Indian Math. Soc. 6(1942) 137-142

[42b] Transformations of arithmetic functions, J. Indian Math. Soc. 6(1942) 143-152

[43a] On arithmetic functions, Proc. Indian Acad. Sci., Sect. A, 18(1943) 88-99

[43b] Identities in multiplicative functions, J. Indian Math. Soc. 7(1943) 48-52

[45] Some generalizations of the divisor function, J. Indian Math. Soc. 9(1945) 32-36

[62] On Vaidyanathaswamy's class division of residue classes modulo N, J. Indian Math. Soc. 26(1962) 167-186

[63] Series associated with Ramanujan's function $\tau(n)$, J. Indian Math. Soc. 27(1963) 57-65
[65] On the sum $\Sigma(a - 1,n)[(a,n) = 1]$, J. Indian Math. Soc. 29(1965) 155-163
[67] An extension of Euler's function, Math. Student 35(1967) 55-59
[72] On functions associated with Vaidyanathaswamy's algebra of classes mod n, Indian J. Pure Appl. Math. 3(1972) 118-141

V. L. Klee
[48] A generalization of Euler's function, Amer. Math. Monthly 55(1948) 358-359

D. M. Kotelyanskiĭ
[53] On N. P. Romanov's method of obtaining identities for arithmetic functions, Ukrain Mat. Žurnal 5(1953) 453-458 (Russian)

K. Krishna
[79] A proof of an identity for multiplicative functions, Canad. Math. Bull. 22(1979) 299-304

J. Lambek
[66] Arithmetic functions and distributivity, Amer. Math. Monthly 73(1966) 969-973

J. Lambek and L. Moser
[55] On integers n relatively prime to f(n), Canad. J. Math. 7(1955) 155-158

E. Langford
[73] Distributivity over the Dirichlet product and complete multiplicative arithmetical functions, Amer. Math. Monthly 80(1973) 411-414

D. H. Lehmer
[30] On the rth divisors of a number, Amer. J. Math. 52(1930) 293-304
[31a] On a theorem of von Sterneck, Bull. Amer. Math. Soc. 37(1931) 723-726
[31b] A new calculus of numerical functions, Amer. J. Math. 53(1931) 832-854
[31c] Arithmetic of double series, Trans. Amer. Math. Soc. 33(1931) 945-957
[36] Polynomials for the n-ary composition of numerical functions, Amer. J. Math. 58(1936) 563-572
[59] Some functions of Ramanujan, Math. Student 27(1959) 105-116

D. N. Lehmer
[13] Certain theorems in the theory of quadratic residues, Amer. Math. Monthly 20(1913) 151-157

K. C. Li
[80] A note on identities for series of type $\Sigma f(n)\mu(n)n^{-s}$, Chinese J. Math. 8(1980) 59-60

A. Liberatore and M. G. M. Tomasini

[79] On two identities of Anderson-Apostol and Landau, La Ricerca 30, No. 3, (1979) 3-10

J. H. Loxton and J. W. Sanders

[80] On an inversion theorem of Möbius, J. Austral. Math. Soc. Ser. A 30(1980) 15-32

U. Maddana Swamy, G. C. Rao and V. Sita Ramajan

[83] On a conjecture in a ring of arithmetic functions, Indian J. Pure Appl. Math. 14(1983) 1519-1530

A. Makowski

[72] Remark on multiplicative functions, Elem. Math. 27(1972) 132-133

J. L. Mauclaire

[76] On the extension of a multiplicative arithmetical function in an algebraic number field, Math. Japon. 21(1976) 337-342

P. J. McCarthy

[58] On a certain family of arithmetic functions, Amer. Math. Monthly 65(1958) 586-590

[59] On an arithmetic function, Monatsh. Math. 63(1959) 228-230

[60a] The generation of arithmetical identities, J. Reine Angew. Math. 203(1960) 55-63

[60b] The probability that (n,f(n)) is r-free, Amer. Math. Monthly 67(1960) 268-269

[60c] Some properties of extended Ramanujan sums, Archiv Math. 11(1960) 253-258

[60d] Some remarks on arithmetical identities, Amer. Math. Monthly 67(1960) 539-548

[60e] Busche-Ramanujan identities, Amer. Math. Monthly 67(1960) 966-970

[62] Some more remarks on arithmetical identities, Portugal. Math. 21(1962) 45-57

[66] Note on some arithmetical sums, Boll. Un. Mat. Ital. 21(1966) 239-242

[68] Regular arithmetical convolutions, Portugal. Math. 27(1968) 1-13

[70] Arithmetical functions and distributivity, Canad. Math. Bull. 13(1970) 491-496

[71] Regular arithmetical convolutions and linear congruences, Colloq. Math. 22(1971) 215-222

[75] The number of restricted solutions of some systems of linear congruences, Rend. Sem. Mat. Univ. Padova 54(1975) 59-68

[77] Counting restricted solutions of a linear congruence, Nieuw. Arch. v. Wisk. 25(1977) 133-147

[86] A generalization of Smith's determinant, Canad. Math. Bull., to appear

A. Mercier

[79] Identité pour $\sum_{\substack{n=1 \\ n\equiv 1(k)}}^{\infty} f(n)/n^s$, Canad. Math. Bull. 22(1979) 317-325

[81a] Remarques sur les représentation de certaines séries de Dirichlet, Canad. Math. Bull. 24(1981) 483-484
[81b] Quelques identités arithmétiques, Ann. Sci. Math. Québec 5(1981) 59-67
[81c] Sommes de la forme $\Sigma g(n)/f(n)$. Canad. Math. Bull. 24(1981) 299-307
[82] Remarques sur les fonctiones spécialement multiplicatives, Ann. Sci. Math. Québec 6(1982) 99-107

J. Morgado
[62] Unitary analogue of the Nagell totient function, Portugal. Math. 21(1962) 221-232: correction, ibid. 22(1963) 119
[63a] Some remarks on the unitary analogue of Nagell's totient function, Portugal. Math. 22(1963) 127-135
[63b] Unitary analogue of a Schemmel's function, Portugal. Math. 22(1963) 215-233
[64] On the arithmetical function σ_h^*, Portugal. Math. 23(1964) 35-40

H. Nadler
[63] Verallgemeinerung einer Formel von Ramanujan, Arch. Math. 14(1963) 243-246

T. Nagell
[23] Verallgemeinerung eines Satzes von Schemmel, Skr. Norske Vod.-Akad. Oslo (Math. Class), I, No. 13(1923) 23-25

K. Nageswara Rao
[61a] On Jordan function and its extension, Math. Student 29(1961) 25-28
[61b] A note on an extension of Euler's ϕ function, Math. Student 29(1961) 33-35
[61c] Generalization of a theorem of Eckford Cohen, Math. Student 29(1961) 83-87
[61d] On extension of Euler's ϕ function, Math. Student 29(1961) 121-126
[61e] On an extension of Ramanujan's sum, Neerajana 1(1961/62) 20-24
[66a] On the unitary analogues of certain totients, Monatsh. Math. 70(1966) 149-154
[66b] An extension of Schemmel's totient, Math. Student 34(1966) 87-92
[66c] Unitary class division of integers (mod n) and related arithmetical identities, J. Indian Math. Soc. 30(1966) 195-205
[67a] On a congruence equation and related arithmetical identities, Monatsh. Math. 71(1967) 24-31
[67b] Some identities involving an extension of Ramanujan's sum, Norske Vid. Selsk. Fohr. (Trondheim) 40(1967) 18-23
[67c] A congruence equation involving the factorisation in residue class ring mod n, Publ. Math. Debrecen, 14(1967) 29-34
[68a] A note on a multiplicative function, Ann. Polon. Math. 21(1968) 29-31
[68b] A ring of arithmetic functions, J. Math. Sci. 3(1968) 31-34

[68c] On an extension of Kronecker's function, Portugal. Math. 27(1968) 169-171
[72] On certain arithmetical sums, Springer Lecture Notes in Math. No. 251, 1972, 181-192

K. Nageswara Rao and R. Sivaramakrishnan
[81] Ramanuman's sum and its applications to some combinatorial problems, Proc. Tenth Manitoba Conf. Numer. Math. and Comput., Vol. II. Congr. Numer. 31(1981) 205-239

W. Narkiewicz
[63] On a class of arithmetical convolutions, Colloq. Math. 10(1963) 81-94

C. A. Nicol
[53] On restricted partitions and a generalization of the Euler ϕ number and the Möbius function, Proc. Nat. Acad. Sci. U.S.A. 39(1963) 963-968
[59] Linear congruences and the Von Sterneck function, Duke Math. J. 26(1959) 193-197
[62] Some formulas involving Ramanujan sums, Canad. J. Math. 14(1962) 284-286
[66] Some diophantine equations involving arithmetic functions, J. Math. Anal. Appl. 15(1966) 154-161

C. A. Nicol and H. S. Vandiver
[54] A von Sterneck arithmetical function and restricted partitions with respect to a modulus, Proc. Nat. Acad. Sci. U.S.A. 40(1954) 825-835: supplement, ibid. 44(1958) 917-918
[55] On generating functions for restricted partitions of rational integers, Proc. Nat. Acad. Sci. U.S.A. 41(1955) 37-42

D. Niebur
[75] A formula for Ramanujan's τ-function, Ill. J. Math. 19(1975) 448-449

J. E. Nymann
[72] On the probability that k-positive integers are relatively prime, J. Number Theory 4(1972) 469-473
[75] On the probability that k-positive integers are relatively prime II, J. Number Theory 7(1975) 406-412

R. P. Pakshirajan
[63] Some properties of the class of arithmetic functions $T_r(N)$, Ann. Polon. Math. 13(1963) 113-114

S. Pankajan
[36] On Euler's ϕ-function and its extensions, J. Indian Math. Soc. 2(1936) 67-75

F. Pellegrino
[53] Sviluppi moderni del calculo numerico integrale di Michele Cipolla, Atti Quarto Cong. Un. Mat. Ital. 2(1953) 161-168

[56] Lineamenti di una teoria delle funzioni aritmetiche, Rend. Mat. Appl. (5) 15(1956) 469-504
[63] La divisione integrale, Rend. Mat. Appl. (5) 22(1963) 489-497
[64] La potenza integrale, Rend. Mat. Appl. (5) 23(1964) 201-220
[66] Operatori lineari dell'anello funzioni aritmetiche, Rend. Mat. Appl. (5) 25(1966) 308-332
[67] Elementi moltiplicativi dell'anello delle funzioni arithmetiche, Rend. Mat. Appl. (5) 26(1967) 422-509

J. Popken
[55] On convolutions in number theory, Proc. Kon. Nederl. Akad. Wetensch. 58(1955) 10-15

M. I. Pulatova
[81] On some systems of linear operators connected with arithmetical inversion formulas, Tsubuka J. Math. 5(1981) 165-172

H. Rademacher
[25] Aufgaben 30-32, Jahr. Deutsch. Math.-Verein. 34(1925), 2. Abteilung, 158-159
[28] Zur additiven Primzahltheorie algebraischer Zahlkorper. III. Über die Darstellung totalpositiver Zahlen als Summen von Totalpositiven Prinzahlen in einem beliebigen Zahlkorper, Math. Zeit. 27(1928) 321-426

K. G. Ramanathan
[43a] On Ramanujan's trigonometrical sum $C_m(n)$, J. Madras Univ. Sect. B 15(1943) 1-9
[43b] Multiplicative arithmetic functions, J. Indian Math. Soc. 7(1943) 111-117
[44] Some applications of Ramanujan's triogonometrical sum $C_m(n)$, Proc. Indian Acad. Sci. (A) 20(1944) 62-69

K. G. Ramanathan and M. V. Subbarao
[80] Some generalizations of Ramanujan's sum, Canad. J. Math. 32(1980) 1250-1260

S. Ramanujan
[16a] Some formulae in the analytic theory of numbers, Messenger Math. 45(1916) 81-84
[16b] On certain arithmetical functions, Trans. Camb. Philos. Soc. 22(1916) 159-184
[18] On certain trigonometrical sums and their applications to the theory of numbers, Trans. Camb. Philos. Soc. 22(1918) 259-276

D. Rearick
[63] A linear congruence with side conditions, Amer. Math. Monthly 70(1963) 837-840
[66a] Semi-multiplicative functions, Duke Math. J. 33(1966) 49-53

[66b] Correlation of semi-multiplicative functions, Duke Math. J. 35(1968) 761-766
[68b] The trigonometry of numbers, Duke Math. J. 35(1968) 767-776

D. Redmond
[79] Some remarks on a paper of A. Ivić, Univ. Beograd Publ. Elektrotehn. Fak. Ser. Mat. Fiz. No. 634-677 (1979) 137-142
[83] A remark on a paper: "An identity involving Ramanujan's sum" [Elem. Math. 36(1981) 16-17] by A. Grytczuk, Elem. Math. 38(1983) 17-20

D. Redmond and R. Sivaramakrishnan
[81] Some properties of specially multiplicative functions, J. Number Theory 13(1981) 210-227

I. M. Richards
[84] A remark on the number of cyclic subgroups of a finite group, Amer. Math. Monthly 91(1984) 571-572

G. J. Rieger
[60] Ramanujansche Summen in algebraishen Zahlkorpern, Math. Nachr. 22(1960) 371-377

R. L. Robinson
[66] An estimate for the enumerative functions of certain sets of integers, Proc. Amer. Math. Soc. 17(1966) 232-237

G.-C. Rota
[64] On the foundations of combinatorial theory I. Theory of Möbius functions, Z. Wahr. 2(1964) 340-368

R. W. Ryden
[73] Groups of arithmetic functions under Dirichlet convolution, Pacific J. Math. 44(1973) 355-360

K. P. R. Sastry
[63] On the generalized type of Möbius function, Math. Student 31(1963) 85-88

U. V. Satyanarayana
[63] On the inversion properties of the Möbius μ-function, Math. Gaz. 47(1963) 38-42
[65] On the inversion properties of the Möbius μ-function II, Math. Gaz. 49(1965) 171-178

U. V. Satyanarayana and K. Pattabhiramasastry
[65] A note on the generalized ϕ-function, Math. Student 33(1965) 81-83

H. Scheid

[68a] Arithmetische Funktionen über Halbordnung, I, J. Reine Angew. Math. 231(1968) 192-214

[68b] Arithmetische Funktionen über Halbordnung, II, J. Reine Angew. Math. 232(1968) 207-220

[69a] Einige Ringe zahlentheoretischer Funktionen, J. Reine Angew. Math. 237(1969) 1-11

[69b] Über ordnungstheoretische Funktionen, J. Reine Angew. Math. 238(1969) 1-13

[70] Funktionen über lokal endlichen Halbordnung, I, Monatsh. Math. 74(1970) 336-347

[71] Funktionen über lokal endlichen Halbordnung, II, Monatsh. Math. 75(1971) 44-56

H. Scheid and R. Sivaramakrishnan

[70] Certain classes of arithmetic functions and the operation of additive convolution, J. Reine Angew. Math. 245(1970) 201-207

W. Schwarz

[76] Aus der Theorie der zahlentheoretischen Funktionen, Jber. Deutsch. Math.-Verein. 78(1976) 147-167

S. L. Segal

[66] Footnote to a formula of Gioia and Subbarao, Canad. Math. Bull. 9(1966) 749-750

L. E. Shader

[70] The unitary Brauer-Rademacher identity, Atti Accad. Naz. Lincei Rend. Cl. Sci. Fis. Mat. Natur. 48(1970) 403-404

H. N. Shapiro

[72] On the convolution ring of arithmetic functions, Comm. Pure Appl. Math. 25(1972) 287-336

J. E. Shockley

[66] On the functional equation $F(mn)F((m,n)) = F(m)F(n)f((m,n))$, Pacific J. Math. 18(1966) 185-189

R. Sitaramachandra Rao and D. Suryanarayana

[70] The number of pairs of integers with L.C.M. $\leq x$, Arch. Math. 21(1970) 490-497

[73] On $\Sigma_{n\leq x}(\sigma^*(n))$ and $\Sigma_{n\leq x}(\phi^*(n))$, Proc. Amer. Math. Soc. 41(1973) 61-66

V. Sita Remaiah

[78] Arithmetical sums in regular convolutions, J. Reine Angew. Math. 303/304(1978) 265-283

[79] On certain multiplicative functions related to the Mobius function, Portugal. Math. 38(1979) 119-134

V. Sita Ramaiah and D. Suryanarayana

[79] Sums of reciprocals of some multiplicative functions, Math. J. Okayama Univ. 21(1979) 155-164

[80] Sums of reciprocals of some multiplicative functions, II, Indian J. Pure Appl. Math. 11(1980) 1334-1355

[81] An order result involving the σ-function, Indian J. Pure Appl Math. 12(1981) 1192-1200

[82a] On a method of Eckford Cohen, Boll. Un. Mat. Ital. 1-B(1982) 1235-1251

[82b] Asymptotic results on sums of some multiplicative functions, Indian J. Pure Appl. Math. 13(1982) 772-784

R. Sivaramakrishnan

[68] The arithmetic function $\tau_{k,r}$, Amer. Math. Monthly 75(1968) 988-989

[69] Generalization of an arithmetic function, J. Indian Math. Soc. 33(1969) 127-132

[70] Problem E2196, Amer. Math. Monthly 77(1970) 772

[71] On three extensions of Pillai's arithmetic function $\beta(n)$, Math. Student 39(1971) 187-190

[74] A number-theoretic identity, Publ. Math. Debrecen 21(1974) 67-69

[76] On a class of multiplicative arithmetic functions, J. Reine Angew. Math. 280(1976) 157-162

[78a] Multiplicative even functions (mod r). I Structure properties, J. Reine Angew. Math. 302(1978) 32-43

[78b] Multiplicative even functions (mod r). II Identities involving Ramanujan sums, J. Reine Angew. Math. 302(1978) 44-50

[79] Square-reduced residue systems (mod r) and related arithmetical functions, Canad. Math. Bull. 22(1979) 207-220

V. Sivaramaprasad and M. V. Subbarao

[85] Regular convolutions and a related Lehmer problem, Nieuw. Arch. v. Wisk. 3(1985) 1-18

A. Sivaramasama

[78] On $\Delta(x,n) = \phi(x,n) - x(\phi(n)/n)$, Math. Student 46(1978) 160-174

D. A. Smith

[67] Incidence functions as generalized arithmetic functions, I. Duke Math. J. 34(1967) 617-634

[69a] Incidence functions as generalized arithmetic functions, II, Duke Math. J. 36(1969) 15-30

[69b] Incidence functions as generalized arithmetic functions, III, Duke Math. J. 36(1969) 353-368

[72] Generalized arithmetic function algebras, Springer Lecture Notes in Math. No. 251, 1972, 205-245

J.-M. Sourian

[44] Géneralisation de certaines formules arithmétiques d'inversion, Revue Sci. (Rev. Rose Illus.) 82(1944) 204-211

H. Stevens

[71] Generalizations of the Euler ϕ function, Duke Math. J. 38(1971) 181-186

M. V. Subbarao

[51] Ramanujan's trigonometrical sum and relative partitions, J. Indian Math. Soc. 15(1951) 57-64

[63] A generating function for a class of arithmetic functions, Amer. Math. Monthly 70(1963) 841-842

[65] The Brauer-Rademacher identity, Amer. Math. Monthly 72(1965) 135-138

[66a] Arithmetic functions satisfying a congruence property, Canad. Math. Bull. 9(1966) 143-146

[66b] A note on the arithmetical functions C(m,r) and C*(n,r), Nieuw Arch. v. Wisk. 14(1966) 237-240

[66c] A congruence for a class of arithmetical functions, Canad. Math. Bull. 9(1966) 571-574

[67a] An arithmetic function and an associated probability theorem, Proc. Kon. Nederl. Akad. Wetensch. 70(1967) 93-95

[67b] A class of arithmetical equations, Nieuw Arch. V. Wisk. 15(1967) 211-217

[68a] Arithmetic functions and distributivity, Amer. Math. Monthly 75(1968) 984-988

[68b] Remarks on a paper of P. Kesava Menon, J. Indian Math. Soc. 32(1968) 317-318

[72] On some arithmetic convolutions, Springer Lecture Notes in Math. No. 251, 1972, 247-271

M. V. Subbarao and Y. K. Feng

[71] On the distribution of the k-free integers in residue classes, Duke Math. J. 38(1971) 741-748

M. V. Subbarao and A. A. Gioia

[67] Identities for multiplicative functions, Canad. Math. Bull. 10(1967) 65-73

M. V. Subbarao and V. C. Harris

[66] A new generalization of Ramanujan's sum, J. London Math. Soc. 4(1966) 595-604

M. V. Subbarao and D. Suryanarayana

[72a] On an identity of Eckford Cohen, Proc. Amer. Math. Soc. 33(1972) 20-24

[72b] Some theorems in additive number theory, Ann. Univ. Sci. Budapest Eotvos Sect. Math. 15(1972) 5-16

[73a] Almost and nearly k-free integers, Indian J. Math. 15(1973) 163-169: Correction and additions, ibid. 17(1975) 172

[73b] The divisor problem for (k,r)-integers, J. Austral. Math. Soc. 15(1973) 430-440

[74] On the order of the error function of the (k,r)-integers, J. Number Theory 6(1974) 112-123
[77] On the order of the error function of the (k,r)-integers, II, Canad. Math. Bull. 20(1977) 397-399
[78] Sums of the divisor and unitary divisor functions, J. Reine Angew. Math. 302(1978) 1-15

P. Subrahmanyan and D. Suryanarayana
[8sa] The maximal square-free, bi-unitary divisor of m which is prime to n. I, Ann. Univ. Sci. Budapest. Eötvös Sect. Math. 25(1982) 163-174
[82b] The maximal square-free, biunitary divisor of m which is prime to n. II, Ann. Univ. Sci. Budapest. Eötvös Sect. Math. 25(1982) 175-192

V. Subrahmanyasastri and J. Hanumanthachari
[79] On some aspects of multiplicative arithmetic functions, Portugal. Math. 38(1979) 193-212

F. Succi
[56] Sulla expressione del quoziente integrale di due funzione aritmetiche, Rend. Mat. Appl. (5) 15(1956) 80-92
[57] Una generalizzazione delle funzioni arithmetiche completamente moltiplicative, Rend. Mat. Appl. (5) 16(1957) 255-280
[60a] Divisibilita integrale localle delle funzione aritmetiche Regolari, Rend. Mat. Appl. (5) 19(1960) 174-192
[60b] Sul gruppo moltiplicativo delle funzioni arithmetiche regolari, Rend. Mat. Appl. (5) 19(1960) 458-472

M. Suganamma
[60] Eckford Cohen's generalizations of Ramanujan's trigonometrical sum C(n,r), Duke Math. J. 27(1960) 323-330

D. Suryanarayana
[67] Asymptotic formula for $\Sigma_{n\leq x}\mu^2(n)/n$, Indian J. Math. 9(1967) 543-545
[68] The number of k-ary divisors of an integer, Monatsh. Math. 72(1968) 445-450
[69a] A generalization of Dedekind's ψ-function, Math. Student 37(1969) 81-86
[69b] The greatest divisor of n which is prime to k, Math. Student 37(1969) 217-218
[69c] Note on Liouville's λ- and Euler's ϕ-functions, Math. Student 37(1969) 217-218
[69d] The number and sum of k-free integers $\leq$ x which are prime to n, Indian J. Math. 11(1969) 131-139
[69e] The number of unitary, squarefree divisors of an integer. I. Norske Vid. Selsk. Forh. (Trondheim) 42(1969) 6-13
[69f] The number of unitary, squarefree divisors of an integer. II, Norske Vid. Selsk. Forh. (Trondheim) 42(1969) 14-21

[70a] Extensions of Dedekind's ψ-function, Math. Scand. 26(1970) 107-118; Addendum, Ibid. 20(1972) 337-338

[70b] A property of the unitary analogue of Ramanujan's sum, Elem. Math. 25(1970) 114

[70c] Uniform O-estimates of certain error functions connected with k-free integers, J. Aurstral. Math. Soc. 11(1970) 242-250: Remark, ibid. 16(1973) 177-178

[71a] New inversion properties of μ^* and μ, Elem. Math. 26(1971) 136-138

[71b] Semi-k-free integers, Elem. Math. 26(1971) 39-40

[71c] Some theorems concerning the k-ary divisors of an integer, Math. Student 39(1971) 384-394

[72a] On the core of an integer, Indian J. Math. 14(1972) 65-74

[72b] The number of bi-unitary divisors of an integer, Springer Lecture Notes in Math., No. 251 (1972) 273-282

[72c] Two arithmetic functions and asymptotic densities of related sets, Portugal. Math. 31(1972) 1-11

[74] On $\Delta(x,n) = \phi(s,n) - x\phi(n)/n$, Proc. Amer. Math. Soc. 44(1974) 17-21

[75] Congruences for sums of powers of primitive roots and Ramanujan's sum, Elem. Math. 30(1975) 129-133; Berichtigung, ibid. 31(1976) 104

[77a] On a theorem of Apostol concerning Möbius functions of order k, Pacific J. Math. 68(1977) 277-281

[77b] On a class of sequences of integers, Amer. Math. Monthly 84(1977) 728-730

[77c] A remark on uniform O-estimate for k-free integers, J. Reine Angew. Math. 298/294(1977) 18-21

[77d] The divisor problem for (k,r)-integers, II, J. Reine Angew. Math. 295(1977) 49-56; correction, ibid. 305(1979) 133

[77e] On the distribution of some generalized square-full integers, Pacific J. Math. 72(1977) 547-555

[78a] On a functional equation relating to the Brauer-Rademacher identity, Enseign. Math. 24(1978) 55-62

[78b] Genralization of two identities of Ramanujan and Eckford Cohen, Boll. Un. Mat. Ital. 15(1978) 424-430

[78c] On the functional equation $s(m,n)F[n/(m,n)] = F(n)h[n/(m,n)]$, Aequationes Math. 18(1978) 322-329

[78d] On the average order of the function $E(x) = \Sigma_{n\le x}\phi(n) - 3x^2/\pi^2$. II. J. Indian Math. Soc. 42(1978) 179-195

[79] On the order of the error function of the square-full integers, Period. Math. Hungar. 10(1979) 261-271

[81] Some more remarks on uniform O-estimate for k-free integers, Indian J. Pure Appl. Math. 12(1981) 1420-1424

[82] On some asymptotic formulae of S. Wigert, Indian J. Math. 24(1982) 81-98

D. Suryanarayana and R. Sitaramachandra Rao

[71] On the order of the error function of the k-free integers, Proc. Amer. Math. Soc. 28(1971) 53-58

[72a] The number of square-full divisors of an integer, Proc. Amer. Math. Soc. 34(1972) 79-80

[72b] On the average order of the function $E(x) = \Sigma_{n\leq x}\phi(n) - 3x^2/\pi^2$, Ark. Mat. 10(1972) 99-106

[73a] The distribution of the square-full integers, Ark. Mat. 11(1973) 195-201

[73b] Uniform O-estimate for k-free integers, J. Reine Angew. Math. 261(1973) 146-152

[73c] On an asymptotic formula of Ramanujan, Math. Scand. 32(1973) 258-264

[73d] Distribution of semi-k-free integers, Proc. Amer. Math. Soc. 37(1973) 340-346

[75a] Distribution of unitarily k-free integers, J. Austral. Math. Soc. 20(1975) 129-141

[75b] The number of bi-unitary divisors of an integer. II. J. Indian Math. Soc. 39(1975) 261-280

[75c] On the true maximum order of a class of arithmetical functions, Math. J. Okayama Univ. 17(1975) 95-100

[76a] The number of unitarily k-free divisors of an integer, J. Austral. Math. Soc. 21(1976) 19-35

[76b] On an identity of Eckford Cohen. II. Indian J. Math. 18(1976) 171-176

D. Suryanarayana and V. Siva Rama Prasad

[70] The number of k-free divisors of an integer, Acta Arith. 17(1970/71) 345-354

[72] The number of pairs of generalized integers with L.C.M. $\leq x$, J. Austral. Math. Soc. 13(1972) 411-416

[73a] Sum functions of k-ary and semi-k-ary divisors, J. Austral. Math. Soc. 15(1973) 148-162

[73b] The number of k-free and k-ary divisors of m which are prime to n, J. Reine Angew. Math. 264(1973) 56-75: Correction, ibid. 265(1974) 182

[74] The number of k-ary divisors of a generalized integer, Portugal. Math. 33(1974) 85-92

[75a] The number of k-ary, (k + 1)-free divisors of an integer, J. Reine Angew. Math. 276(1975) 15-35

[75b] The number of k-free divisors of an integer. II. J. Reine Angew. Math. 276(1975) 200-205

[77] The number of semi-k-free divisors of an integer, Ann. Univ. Sci. Budapest Eötvös Sect. Math. 20(1977) 5-19

D. Suryanarayana and M. V. Subbarao

[80] Arithmetical functions associated with the bi-unitary k-ary divisors of an integer, Indian J. Math. 22(1980) 281-298

D. Suryanarayana and P. Subrahamanyan

[77] The maximal k-free divisor of m which is prime to n. I. Acta Math. Acad. Sci. Hungar. 30(1977) 49-67

[79] The maximal k-free divisor of m which is prime to n. II. Acta Math. Acad. Sci. Hungar. 33(1979) 239-260
[81a] The maximal k-full divisor of an integer, Indian J. Pure Appl. Math. 12(1981) 175-190
[81b] The maximal k-free unitary divisor of an integer, Bull. Inst. Math. Acad. Sinica 9(1981) 472-488

D. Suryanarayana and D. T. Walker
[77] Some generalizations of an identity of Subhankulov, Canad. Math. Bull. 20(1977) 489-494

O. Szász
[47] On Möbius inversion formula and closed sets of functions, Trans. Amer. Math. Soc. 62(1947) 213-239

P. Szüsz
[67] Once more the Brauer-Rademacher theorem, Amer. Math. Monthly 74(1967) 570-571

P. Thrimurthy
[78] Some identities in arithmetic functions, Math. Student 46(1978) 251-258

R. Vaidyanathaswamy
[27] On the inversion of multiplicative arithmetic functions, J. Indian Math. Soc., Notes & Questions 17(1927) 69-73
[30] The identical equations of the multiplicative function, Bull. Amer. Math. Soc. 33(1931) 579-662
[31] The theory of multiplicative arithmetic functions, Trans. Amer. Math. Soc. 33(1931) 579-662
[37] A remarkable property of integers mod N and its bearing on group theory, Proc. Indian Acad. Sci. Sect. A 5(1937) 63-75

A. C. Vasu
[65] A generalization of Brauer-Rademacher identity, Math. Student 33(1965) 97-101
[66] On a certain arithmetic function, Math. Student 34(1966) 93-95
[72a] A generating function for $J_{k,1}(m)$, Math. Student 40(1972) 255-259
[72b] A note on an extension of Klee's ψ-function, Math. Student 40A(1972) 36-39
[73a] Generating functions for $J_{k,1}(m)$, $V_{k,1}(m)$, $\lambda'(m,n)$ and $I^2(m)$, Math. Student 41(1973) 376-380
[73b] On three product functions, Math. Student 41(1973) 381-384

S. Venkatramaiah
[73] On a paper of Kesava Menon, Math. Student 41(1973) 303-306

C. S. Venkataraman

[46a] A new identical equation for multiplicative functions of two arguments and its application to Ramanujan's sum $C_M(N)$, Proc. Indian Acad. Sci. Sect. A 24(1946) 518-529

[46b] On some remarkable types of multiplicative functions, J. Indian Math. Soc. 10(1946) 1-12

[46c] Further applications of the identical equation to Ramanujan's sum $C_M(N)$ and Kronecker's function $\rho(M,N)$, J. Indian Math. Soc. 10(1946) 57-61

[46d] The ordinal correspondence and certain classes of multiplicative functions of two arguments, J. Indian Math. Soc. 10(1946) 81-101

[46e] A theorem on residues and its bearing on multiplicative functions with a modulus, Math. Student 14(1946) 59-62

[49a] Classification of multiplicative functions of two arguments based on the identical equation, J. Indian Math. Soc. 13(1949) 17-22

[49b] On Von Sterneck-Ramanujan function, J. Indian Math. Soc. 13(1949) 65-72

[49c] A generalization of Euler's ϕ-function, Math. Student 17(1949) 34-36

[50] Modular multiplicative functions, J. Madras Univ. Sect. B. 19(1950) 69-78

C. S. Venkataraman and R. Sivaramakrishnan

[72] A extension of Ramanujan's sum, Math. Student 40A(1972) 211-216

T. Venkataraman

[74] A note on the generalization of an arithmetic function in kth power residues, Math. Student 42(1974) 101-102

L. Vietoris

[67] Über die Zahl der in einem k-reduzierten Restsystem liegenden Losungen einer Kongruenz $x_1 + x_2 + \ldots + x_r \equiv a(m^k)$, Monatsh. Math. 71(1967) 55-63

[68] Über eine Zahlfunktion von K. Nagaswara Rao, Monatsh. Math. 72(1968) 147-151

M. Ward

[39] The algebra of lattice functions, Duke Math. J. 5(1939) 357-371

L. Weisner

[35a] Abstract theory of inversion of finite series, Trans. Amer. Math. Soc. 38(1935) 474-484

[35b] Some properties of prime-power groups, Trans. Amer. Math. Soc. 38(1935) 485-492

S. Wigert

[32] Sur quelques formules asymptotiques de la theorie des nombres, Ark. Mat. Astron. Fys. 22B(1932) 1-6

R. Wiegandt
[59] On the general theory of Möbius inversion formula and Möbius product, Acta Sci. Math. Szeged 20(1959) 164-180

B. M. Wilson
[23] Proofs of some formulae enunciated by Ramanujan, Proc. London Math. Soc. 21(1923) 235-255

K. Wohlfarht
[57] Über Operatoren Heckesche Art bei Modulformen reeler Dimension, Math. Nachr. 16(1957) 233-256
[73] Über Funktionalgleichungen zahlentheoretischer Funktionen, Colloq. Math. 27(1973) 278-281

K. L. Yocom
[73] Totally multiplicative functions in regular convolution rings, Canad. Math. Bull. 16(1973) 119-128

Index

A-core function γ_A 166
A-even function (mod r) 170
A-multiplicative function 172
A-specially multiplicative
  function 174
abscissa of absolute con-
  vergence 185
Anderson, D. R. 111, 252
Apostol, T. M. 63, 64, 66, 111,
  113, 250, 252
arithmetical functions 1
  B-convolution of 179
  Bell series of 60
  Cauchy product of 76
  cross between 53
  Dirichlet convolution of 2
  Dirichlet series of 184
  generalized 294
  generating function of 187
  inverse of 4
  K-convolution of 149
  kth convolute of 53
  product of 2
  regular arithmetical con-
    volution of 155
  ring of 4
  sum of 2
  summatory function of 262
asymptotic formula 262

B-convolution 179
B-multiplicative function 179
basic sequence 178
Bell, E. T. 62, 69
Bell series 60
Bender, E. A. 330, 332
Bernoulli, D. 62
Beumer, M. G. 69
van der Blij, F. 68
block factor 52
Brauer, A. 111, 146
Brauer-Rademacher identity 75
  generalized 88
Busche-Ramanujan identity 25
  restricted 52

Carlitz, L. 63, 64, 331
Carmichael, R. D. 112
Carroll, T. B. 65
Cashwell, E. 63
Cauchy product 76
chain 297
Chidambaraswamy, J. 68
Chowla, S. 66, 251
Christopher, J. 292
Cipolla, M. 62
Cohen, E. 63, 64, 68, 69,
  111, 112, 146, 147, 148,
  181, 182, 252, 291, 330
companion function 312
completely multiplicative
  function 16
constant on chains 312
convolution
  B- 179
  Dirichlet 2
  K- 149
  of a generalized arith-
    metical function 294
  regular arithmetical 155
  unitary 155
core function 31
van der Corput, J. G. 292
Crum, M. M. 252
cyclotomic polynomial 29

Davison, T. M. K. 181, 182
Daykin, D. E. 63
Dedekind's function ψ 41
Deligne, P. 68
derangement 328
Dickson, L. E. 62, 64, 69,
  113, 152
Dirichlet series 184
distributive lattice 299
Dixon, J. D. 146

Euler, L. 62
Euler function ϕ 1
Euler product 190
Euler's constant 256

Euler Summation Formula 258
Everett, C. 63
Ewell, J. A. 68
even function (mod r) 79

factorable incidence
 function 304
Fantino, I. P. 182
formal power series 58
Fourier coefficients 82
Fredman, M. L. 182
function
 A-multiplicative 172
 A-specially multipli-
 cative 174
 B-multiplicative 179
 arithmetical 1
 Cauchy product of 76
 companion 312
 completely multipli-
 cative 16
 even (mod r) 79
 generalized arith-
 metical 294
 incidence 294
 linear 64
 multiplicative 7
 periodic (mod r) 76
 quadratic 65
 (r,k)-even 105
 Ramanujan tau 66
 Riemann zeta 185
 specially multipli-
 cative 24
 totient 53
 β, β_k 25, 51
 γ, the core 31
 γ_A, the A-core 166
 δ 4
 δ_k 34
 ζ, the zeta 2
 ζ_k 2
 $\zeta_{k,q}$ 57
 θ, θ_k 36, 37
 $\theta_{k,q}$ 58
 $\theta(\cdot,r)$, the Nagell 119
 λ, Liouville's 45
 $\lambda_{k,q}$ 56
 Λ, Mangoldt's 247
 μ, the Möbius 5
 μ_k 39
 $\bar{\mu}_k$ 233
 $\mu_{k,q}$ 57
 μ_A 155
 ν_s 55
 ξ_k 228
 $\rho_{k,s}$, Gegenbauer's 55
 σ, σ_k 2, 1
 σ_k' 235
 τ 2
 τ_k, $\tau_{k,q}$ 40
 ϕ, the Euler 1
 ϕ_k 42
 $\phi_{k,q}$ 58
 $\phi(\cdot,1)$, Schemmel's 35
 ϕ_A, ϕ_U 162, 273
 Φ_k, Klee's 38
 χ 27
 ψ, Dedekind's 41
 ψ_k 41
 Ψ_k 42
 ω 33
 H_k, the von Sterneck 14
 J_k, Jordan's 13

q_k 41
R, R_1 26, 27

Gagliardo, E. 111
Gegenbauer, L. 252
Gegenbauer's function $\rho_{k,s}$ 55
generalized arithmetical
function 294
generalized Ramanujan sum 104
generating function 187
Gessley, M. D. 182
Gioia, A. A. 65, 181, 182, 183
Goldberg, R. R. 254
Goldman, J. R. 330, 332
Goldsmith, D. L. 183, 251
greatest lower bound 298

Hall, P. 330
Hansen, R. T. 63
Hanumanthachari, J. 69, 251
Hardy, G. H. 68, 253
Harris, V. C. 69, 147, 251, 252
Haviland, E. D. 69, 212
Hecke, E. 66
Hille, E. 254
Hölder, O. 111

identical equation 65
incidence functions 294
convolution of 294
factorable 304
inverse of 295
product of 294
ring of 295
sum of 294
incidence ring 295
Inclusion-Exclusion Principle 12
interval 293

Jordan function J_k 13

K-convolution 149
k-free part 56
k-residue system (mod n) 44
(k,q)-integers 56
kth convolute 53
Kesava Menon, P. 68, 111
Klee, V. 68, 69
Klee's function Φ_k 38
Kotelyanskiĭ, D. M. 251
Krishna, K. 183

Lambek, J. 64, 65, 251
Landau, E. 250, 252
Langford, E. 64
lattice 298
distributive 299
local 302
lower semi- 302
modular 320
semimodular 320
least upper bound 298
Lehmer, D. H. 65
Lehmer, D. N. 147
linear function 64
Liouville, J. 62
Liouville's function λ 45
local lattice 302
locally distributive 302
locally modular 302
locally finite poset 293
lower semi-lattice 302
Loxton, J. H. 63, 254

Mangoldt's function Λ 247
McCarthy, P. J. 63, 66, 69,
112, 113, 147, 182, 292, 331
mean value 272
Mercier, A. 252
Möbius, A. F. 63, 253, 254
Möbius function μ 5, 296
Möbius inversion formula 5
modular lattice 320
Morgado, J. 182
multiplicative function 7
norm of 50

(n,k)-residue system 42
Nagell, T. 146
Nagell function $\theta(\cdot,r)$ 119
Nageswara Rao, R. 68, 69,
113, 147
Narkiewicz, W. 182
Nicol, C. A. 111, 146
nonprincipal character
(mod 4) 27

norm 50
numbers of solutions
    $\theta(n,r)$ 119
    $\theta(m,n,r)$ 135
    $M(n,r,s)$ 121
    $M(n_1,\ldots,n_t,r,s)$ 132
    $N(n,r,s)$ 116
    $N_k(n,r,s)$ 126
    $N'(n,r,s)$ 127
    $N_A(n,r,s)$ 167
    $N(n_1,\ldots,n_t,n,s)$ 135
    $P_k(n,r,s)$ 125
    $Q_k(n,r,s)$ 126
    $S(n,r,s)$ 128
Pankajan, S. 64
partially ordered set 293
Pattabhiramasastry, K. 69, 251
Pellegrino, F. 62
periodic (mod r) 76
poset 293
    locally finite 293
primitive integer 160

quadratic function 65

(r,k)-even function 105
Rademacher, H. 111, 113, 146
Rademacher sums 113
Ramanathan, K. G. 67, 112, 113, 146
Ramanujan, S. 66, 67, 68, 252
Ramanujan sum 70
    generalized 104
Ramanujan tau-function 66
Rearick, D. 146, 254
Redmond, D. 251
regular arithmetical convolution 155
restricted Busche-Ramanujan identity 52
Richards, I. M. 68
Rieger, G. J. 113
Riemann zeta function 185
ring of arithmetical functions 4
ring of generalized arithmetical functions 295

Rogel, F. 69
Rota, J.-C. 330, 332

Sanders, J. W. 63, 254
Satyanarayana, U. V. 63, 69, 251
Scheid, H. 66, 331, 332
Schemmel's function $\phi(\cdot,1)$ 35
Schwarz, W. 292
semi-linear congruence 128
semimodular lattice 320
Shapiro, H. N. 63
Sita Ramaiah, V. 68
Sivaramakrishnan, R. 64, 65, 66, 68, 69, 113, 251
Smith, D. 330, 331, 332
Smith, H. J. S. 113
Smith's determinant 33, 113
solution of a congruence 114
specially multiplicative function 24
Stevens, H. 68
von Sterneck, R. D. 64, 111
von Sterneck function H_k 14
von Sterneck number 111
Subbarao, M. V. 63, 69, 111, 113, 147, 181, 182, 183, 251, 252
Suganamma, M. 112, 147
summatory function 262
sums
    $B(n,r)$ 108
    $c(n,r)$, Ramanujan 70
    $c_k(n,r)$, generalized Ramanujan 104
    $c_A(n,r)$ 164
    $c(n_1,\ldots,n_t,r)$ 131
    $c^{(k)}(n,r)$ 107
    $D_{k,q}(n,r)$ 138
Suryanarayana, D. 69, 113, 251, 291
Swanson, L. G. 63
Szász, O. 254
Szüsz, P. 111

totally even function (mod r) 131
totally ordered set 297

totient 53
type 159

unitary convolution 155
unitary divisor 52, 154

Vaidyanathaswamy, R. 62, 64, 65, 66, 69, 182, 183
Vandiver, H. S. 111, 146
Varga, R. S. 254
Vasu, A. C. 69, 111
Venkataraman, T. 68
Venkatramaiah, S. 68
Vietoris, L. 146, 147

Ward, M. 330
Weisner, L. 330, 332
Wiegandt, R. 330
Wigert, S. 69, 251
Wilson, B. M. 251
Wright, E. M. 253

Yocum, K. L. 183

zeta function 2